VOLUME ONE HUNDRED AND FIFTY SEVEN

Current Topics in
DEVELOPMENTAL BIOLOGY
Organizers in Development

CURRENT TOPICS IN DEVELOPMENTAL BIOLOGY

"A meeting-ground for critical review and discussion of developmental processes"
A.A. Moscona and Alberto Monroy (Volume 1, 1966)

SERIES EDITOR

Paul M. Wassarman
Department of Cell, Developmental and Regenerative Biology
Icahn School of Medicine at Mount Sinai
New York, NY, USA

CURRENT ADVISORY BOARD

Blanche Capel
Denis Duboule
Anne Ephrussi
Susan Mango

Philippe Soriano
Claudio Stern
Cliff Tabin
Magdalena Zernicka-Goetz

FOUNDING EDITORS

A.A. Moscona and Alberto Monroy

FOUNDING ADVISORY BOARD

Vincent G. Allfrey
Jean Brachet
Seymour S. Cohen
Bernard D. Davis
James D. Ebert
Mac V. Edds, Jr.

Dame Honor B. Fell
John C. Kendrew
S. Spiegelman
Hewson W. Swift
E.N. Willmer
Etienne Wolff

VOLUME ONE HUNDRED AND FIFTY SEVEN

CURRENT TOPICS IN
DEVELOPMENTAL BIOLOGY

Organizers in Development

Edited by

CLAUDIO D. STERN
Department of Cell & Developmental Biology,
University College London, Gower Street,
London, WC1E 6BT, United Kingdom

Academic Press is an imprint of Elsevier
125 London Wall, London, EC2Y 5AS, United Kingdom
50 Hampshire Street, 5th Floor, Cambridge, MA 02139, United States
525 B Street, Suite 1650, San Diego, CA 92101, United States

First edition 2024

Copyright © 2024 Elsevier Inc. All rights are reserved, including those for text and data mining, AI training, and similar technologies.

Publisher's note: Elsevier takes a neutral position with respect to territorial disputes or jurisdictional claims in its published content, including in maps and institutional affiliations.

No part of this publication may be reproduced or transmitted in any form or by any means, electronic or mechanical, including photocopying, recording, or any information storage and retrieval system, without permission in writing from the publisher. Details on how to seek permission, further information about the Publisher's permissions policies and our arrangements with organizations such as the Copyright Clearance Center and the Copyright Licensing Agency, can be found at our website: www.elsevier.com/permissions.

This book and the individual contributions contained in it are protected under copyright by the Publisher (other than as may be noted herein).

Notices
Knowledge and best practice in this field are constantly changing. As new research and experience broaden our understanding, changes in research methods, professional practices, or medical treatment may become necessary.

Practitioners and researchers must always rely on their own experience and knowledge in evaluating and using any information, methods, compounds, or experiments described herein. In using such information or methods they should be mindful of their own safety and the safety of others, including parties for whom they have a professional responsibility.

To the fullest extent of the law, neither the Publisher nor the authors, contributors, or editors, assume any liability for any injury and/or damage to persons or property as a matter of products liability, negligence or otherwise, or from any use or operation of any methods, products, instructions, or ideas contained in the material herein.

ISBN: 978-0-12-823759-5
ISSN: 0070-2153

For information on all Academic Press publications
visit our website at https://www.elsevier.com/books-and-journals

Publisher: Zoe Kruze
Editorial Project Manager: Devwart Chauhan
Production Project Manager: Kumaresan Chandrakumar
Cover Designer: Miles Hitchen
Typeset by MPS Limited, India

Contents

Contributors ix

Preface: Hans Spemann, Hilde Mangold and the 'organizer' xi

1. The organizer: What it meant, and still means, to developmental biology   **1**
Jonathan Slack

  1. What did Spemann and his colleagues believe in 1921–22?   2
  2. What was the view in the 1930s?   10
  3. Evocation and individuation   12
  4. Neural and mesodermal inducers   16
  5. The two gradient models   17
  6. Later results   19
  7. Mesoderm induction   20
  8. The re-emergence of experimental embryology in a "modern" guise   23
  9. The second gold rush   25
      9.1 Polarization of the fertilized egg   26
      9.2 Mesoderm induction   28
      9.3 Neuralization/dorsalization   30
      9.4 Antero-posterior patterning   32
  10. How we should understand the organizer today   34
  Acknowledgments   35
  References   35

2. The organizer and neural induction in birds and mammals   **43**
Claudio D. Stern

  1. Historical introduction   44
  2. Hensen's node ("the node")   44
      2.1 Cellular composition of the node   45
  3. Neural induction: spatial and temporal aspects   47
  4. Molecular mechanisms of neural induction   49
  5. The organizer and "dorso-ventral" (axial-lateral) patterning   53
  6. Neural induction and rostro-caudal (anterior-posterior) patterning: how many organizers?   53

7. Is neural induction by a grafted organizer comparable to normal neural plate development?	56
8. Neural induction in vivo and in vitro	58
9. Are the mechanisms of neural induction in amniotes different from those in anamniotes?	58
References	60

3. Tissues and signals with true organizer properties in craniofacial development — 67
Shruti S. Tophkhane and Joy M. Richman

1. Introduction	68
2. Neural crest cells contain craniofacial patterning information, but do they have organizer properties?	69
3. Foregut endoderm is an organizer in facial patterning	69
4. Nasal placodes as craniofacial organizers	74
5. Frontonasal mass epithelial zone a potential facial organizer	75
6. Signals with craniofacial organizer properties	75
6.1 Sonic Hedgehog and Fibroblast growth factor	75
6.2 Noggin and retinoic acid	77
6.3 Endothelin	78
7. Concluding remarks	79
Acknowledgments	79
References	80

4. Organizing activities of axial mesoderm — 83
Elizabeth Manning and Marysia Placzek

1. The embryonic organizer, neural induction and secondary organizers	84
2. The role of axial mesoderm in the chick	87
2.1 Chick axial mesoderm development	87
2.2 Young head process mesoderm and its derivatives stabilize anterior neural fate	88
2.3 Chick prechordal mesoderm/mesendoderm acts as a local organizer along the A-P and D-V axes	91
2.4 Young head process mesoderm promotes morphogenesis	92
2.5 Chick notochord acts as a local organizer, potentially influencing all three axes	93
3. The role of axial mesoderm in Xenopus and zebrafish	95
3.1 Axial mesoderm development in Xenopus and zebrafish	95

    3.2 In Xenopus and zebrafish, prechordal mesoderm/mesendoderm maintains A-P regional identity   96
    3.3 Prechordal mesoderm/mesendoderm and notochord reveal organization through their impact on morphogenesis   101
    3.4 In Xenopus and zebrafish, prechordal mesoderm/mesendoderm and notochord maintain regional identity along the D-V and medio-lateral axes   103
 4. The role of axial mesoderm in mouse   104
    4.1 Mouse axial mesoderm development   104
    4.2 Mouse prechordal mesoderm/mesendoderm and notochord act as local organizers   104
    4.3 Evolutionarily-conserved secreted factors mediate PMe and notochord activities   106
    4.4 Prechordal mesoderm/mesendoderm-derived signaling factors   107
    4.5 Notochord-derived signaling factors   109
 5. Summary   112
References   112

5. Transport and gradient formation of Wnt and Fgf in the early zebrafish gastrula   **125**

Emma J. Cooper and Steffen Scholpp

1. Introduction   126
2. The discovery of the Spemann-Mangold Organiser   127
3. Morphogen signalling from the organiser   128
4. Comparing the Wnt and Fgf signalling pathways on a molecular level   129
5. Contrasting the role of Wnt and Fgf signalling within embryonic patterning   130
    5.1 Neural induction   130
6. Anteroposterior axis formation   131
7. Neural AP axis patterning   132
8. Mesodermal and endodermal fate   133
9. Juxtaposing the transport mechanisms for Wnt and Fgf   134
10. Post-translational modification   134
11. Carrier proteins   135
12. Restricted diffusion by heparan sulphate proteoglycans (HSPGs)   135
13. Extracellular vesicles   138
14. Cytonemes facilitate paracrine morphogen signalling   139
15. Comparing Wnt and Fgf signalling gradient formation   141
    15.1 Gradients through control and clearance   141

16.	Controlled morphogen transport shapes the gradient	142
17.	Signalling modulators in the target cells	143
18.	Final remarks on the importance of transport modes in morphogen gradient formation	144
Acknowledgement		146
References		146

Contributors

Emma J. Cooper
Living Systems Institute, Faculty of Health and Life Sciences, University of Exeter, Exeter, United Kingdom

Elizabeth Manning
School of Biosciences; Bateson Centre; University of Sheffield, Sheffield, United Kingdom

Marysia Placzek
School of Biosciences; Bateson Centre; Neuroscience Institute, University of Sheffield, Sheffield, United Kingdom

Joy M. Richman
Life Sciences Institute and Faculty of Dentistry, University of British Columbia, Vancouver, BC, Canada

Steffen Scholpp
Living Systems Institute, Faculty of Health and Life Sciences, University of Exeter, Exeter, United Kingdom

Jonathan Slack
Department of Life Sciences, University of Bath, Bath, United Kingdom

Claudio D. Stern
Department of Cell and Developmental Biology, University College London, London, United Kingdom

Shruti S. Tophkhane
Life Sciences Institute and Faculty of Dentistry, University of British Columbia, Vancouver, BC, Canada

Preface: Hans Spemann, Hilde Mangold and the 'organizer'

The year 2024 marks the 100th anniversary of the publication of one of the most influential papers in the history of Developmental Biology: the first clear account of the existence of an "organizer" by Hans Spemann (1869-1941) and Hilde Mangold (1898-1924) (Spemann and Mangold, 1924). Spemann worked for his PhD with Theodor Boveri, studying on cell lineages in a nematode worm (*Strongylus paradoxus*). Later he turned to regeneration of parts of the eye in the newt; among several discoveries, he reported that the lens was induced to form from the head ectoderm under the influence of the optic vesicle, an outgrowth from the forebrain that later gives rise to the retina and other eye structures (Spemann, 1901). The idea of induction, as an interaction between one tissue that acts as the source of signals (the 'inducer') and a receiving tissue that is able to respond to the signal by changing its direction of differentiation (Gurdon, 1987) was by then already established, particularly from the work of Curt Herbst (Oppenheimer, 1991), but even the idea that the lens was the result of an interaction between tissues had already been proposed by von Baer as early as 1828 (Von Baer, 1828; Oppenheimer, 1991).

Spemann then turned his attention to the problem of whether early development is a 'regulative' process – then a subject of intense debate following apparently contradictory experiments of Wilhelm Roux and Hans Driesch. Roux had found that killing one of the two blastomeres following the first cell division of a frog embryo resulted in the formation of only the left- or the right-half of the gastrula stage embryo, which was interpreted to mean that cell identities are already fixed at the earliest stages of development, perhaps even in the egg ('mosaic' development) (Roux, 1888). A few years later, Driesch, experimenting on sea urchin embryos, carefully separated the first two cells, finding that both were able to give rise to a complete embryo, therefore suggesting that development is 'regulative' – cells determine their identities at least in part as a result of interaction with their neighbors (Driesch, 1892). A few years later, Thomas Hunt Morgan had validated Driesch's conclusions in a frog embryo, showing that the discrepancy between the earlier studies was not due to the different species used (Morgan, 1895). Spemann extended these observations using newt embryos at different stages of development; armed with a fine baby's hair (reportedly that of his daughter Margrette) tied into a knot

(Sander and Faessler, 2001), he isolated cells (and also different fragments of the fertilized egg) at different stages of development of a newt embryo, concluding that indeed some portions of the embryo can give rise to an entire embryo, as well as establishing that the relationship between the isolated fragments and the plane of cleavage is important, as is the inclusion of the "grey crescent" (future dorsal side of the embryo) in fragments that will generate the entire embryo (reviewed by Spemann, 1938).

During this exciting period in the history of Developmental Biology, an American embryologist, Warren H. Lewis, was performing transplants of different parts of amphibian embryos to different locations. One of his experiments involved transplantation of the dorsal lip of the blastopore, which in some cases led to the formation of an ectopic (extra) neural plate – Lewis suggested that the reason for this was that the dorsal lip of the blastopore contains the precursor cells that will give rise to the nervous system, which self-differentiate when placed elsewhere in the host embryo (Lewis, 1907). Because of the background of studies on cell interactions in the lens and between blastomeres discussed above, Spemann wondered whether an alternative interpretation of this experiment could instead be that the graft had induced the extra neural plate, rather than self-differentiated. Indeed he had already suggested that the neural plate might be the result of an inductive interaction with the archenteron roof (which is derived from the dorsal lip of the blastopore) as early as 1903 (Spemann, 1903). So he started to repeat Lewis's experiments, but quickly realized that the definitive answer to the question required the donor and host tissues to be labelled in some way so that they could be distinguished.

Continuing his interest in regeneration and cell interactions, Spemann's attention had been drawn to experiments conducted in the USA by G. Ethel Browne on the cnidarian *Hydra*, which can regenerate both its foot and its head/tentacles when these are severed. Ethel Browne discovered that grafts of the hypostome, tissue situated just below the head of the hydra, could generate ectopic outgrowths and a whole new organism. Using pigmented (*Hydra viridissima*) and unpigmented species, she ascertained that the outgrowth was at least partly the result of an inductive interaction rather than self-differentiation of the graft (Browne, 1909). Browne had sent a signed reprint of her paper to Spemann two months after its publication (Lenhoff, 1991) – it obviously made a mark in Spemann's thinking. In 1919/1920, a new student arrived in the lab, Hilde Proescholdt. Spemann charged Hilde to attempt to reproduce Ethel

Browne's findings in Hydra. For about a year, Hilde performed these experiments and kept clear notes, confirming that indeed there appeared to be an inductive interaction. The original aim of Hilde's project had been to use this technology to repeat experiments done by Abraham Trembley in the 18th Century in which Hydra animals were turned inside out, placing the endoderm outermost – however Hilde did not succeed in doing this (Hamburger, 1988, Lenhoff, 1991, Sander and Faessler, 2001). But the combination of circumstances inevitably led to the question of whether differently colored species of newt embryo could be used to distinguish between the contributions of graft and host tissue in Warren Lewis's dorsal lip transplantation experiments. They resorted to three species of newt: *Triton (Triturus) cristatus* (the great crested newt), *T. taeniatus* and *T. alpestris* – the latter two are darkly pigmented, the first one more lightly colored. In 1921, Spemann published a paper with the first results of these experiments – as sole author (Spemann, 1921). Hilde had by then performed a few interspecies transplants and the first successful one was her 8th attempt, Um.8, which had a clear ectopic neural plate on the ventral side and the light colored cells (derived from the transplant) were confined to the midline, a notochord like structure. Spemann added this result only as a footnote in the 1921 paper and Hilde was not mentioned at all!

By 1924, a year after Hilde completed her PhD, she had performed more than 200 grafts but only a few survived (about 5 with clear results) because of infections or other problems. The famous paper appeared soon after Hilde died – different versions of the cause of her death exist, the most widely cited is that it was the result of an explosion of a domestic heater, and another is that she committed suicide after suffering depression following the birth of her son Christian. The father was Otto Mangold, Spemann's laboratory assistant at the time, whom Hilde married soon after the start of her PhD (Spemann, 1938, Mangold, 1953, Hamburger, 1988) (and Salomé Gluecksohn-Waelsch, personal communication c. 1993). Hans Spemann was the sole awardee of the Nobel prize for Physiology or Medicine in 1935. He is reported to have ended his acceptance speech with a Nazi salute, although it was Otto Mangold who was the more fervent member of the Nazi party during the War.

From this brief historical context, about which a considerable amount has been written, it is obvious that the famous 1924 paper does not stand alone in a vacuum but is the result of numerous earlier contributions, both by the main author and by many others, some dating back almost a century before that. But it remains a landmark, not only by the elegance of the experimental design and the skill of the experiments themselves and the very careful reporting of the results, but also by the integration of ideas from previous studies in a concise and extremely clear way. The main innovation of the paper was not the concept of induction, or even of 'neural induction', but rather the observation that a single tissue could convey both a change in the direction of differentiation (induction) and a coherent organization of the responding tissue into head-tail structures and appropriate recruitment of associated tissues into a coherent embryonic axis. This observation is what led to the idea of the "organizer" as an orchestrator of developmental decisions (Anderson and Stern, 2016).

This issue contains 5 vignettes into some of the legacy of the 1924 paper. First, a review by Jonathan Slack summarizing our knowledge of the organizer and neural induction mainly based on studies in the anuran *Xenopus laevis* and other non-amniote embryos (Slack, 2024). Then, a review of progress made on the organizer and neural induction in amniote (bird and mammalian) embryos (Stern, 2024). Tophkhane and Richman then write about tissues that could act as organizers during craniofacial development in vertebrates (Thophkhane and Richman, 2024) and Manning and Placzek survey the evidence for inducing and patterning (organizer) functions of the axial mesoderm, tissues derived from the original ('primary', or 'Spemann-Mangold') organizer (Manning and Placzek, 2024). One long-standing question about how organizers work is how the signal(s) they emit is/are transmitted to reach the target responding tissues, and how the range of this transmission is regulated. These are the questions touched upon by the final contribution (Cooper and Scholpp, 2024).

One hundred years on, we have learned quite a lot, but many key questions remain for the future. Among them, one that has hardly been addressed includes whether the organizer represents a uniform signaling population of cells that emit an inducing and patterning substance(s), or whether it is a collection of cells with distinct properties that define the organizer activity collectively, through the sum of their secreted products. Another key question is "what is a developmental decision?" – at what point after receiving signals from the organizer can the responding tissue be considered to have changed its direction of differentiation (ie its fate), and what happens in the intervening time? Yet another is: "how many different organizers are involved in patterning the organism?". At this point we are only just starting to make inroads into these fundamental questions, and much remains to be done before we really understand organizers and how they work, as well as when an organizer, rather than just a simple inducing interaction between cells, is essential for generating complexity.

Claudio D. Stern
Department of Cell & Developmental Biology,
University College London, Gower Street,
London, WC1E 6BT, United Kingdom

References

Anderson, C., & Stern, C. D. (2016). Organizers in Development. *Curr. Top. Dev. Biol. 117*, 435–454.

Browne, G. E. (1909). The production of new hydranths in hydra by the insertion of small grafts. *J exp Zool, 7*, 1–37.

Cooper, E. J., & Scholpp, S. (2024). Transport and gradient formation of Wnt and Fgf in the early zebrafish gastrula. *Curr. Top. Dev. Biol. 157*, 125–154.

Driesch, H. (1892). Entwicklungsmechanische Studien: I. Der Werth der beiden ersten Furschungszellen in der Echinodermenentwicklung. *Experimentelle Erzeugen von Teil und Doppelbildung. Z f. wiss. Zool. 53*, 160–178.

Gurdon, J. B. (1987). Embryonic induction - molecular prospects. *Development, 99*, 285–306.

Hamburger, V. (1988). *The heritage of experimental embryology: Hans Spemann and the Organizer*. Oxford: Oxford University Press.

Lenhoff, H. M. (1991). Ethel Browne, Hans Spemann, and the discovery of the Organizer phenomenon. *Biol Bull, 181*, 72–80.

Lewis, W. H. (1907). Transplantation of the lips of the blastopore in Rana pipiens. *Am J Anat, 7*, 137–141.

Mangold, O. (1953). *Hans Spemann: ein Meister der Entwicklungsphysiologie, sein Leben und sein Werk*. Stuttgart: Wissenschaftliche Verlagsgesellschaft.

Manning, E., & Placzek, M. (2024). Organizing activities of axial mesoderm. *Curr. Top. Dev. Biol. 157*, 83–124.

Morgan, T. H. (1895). Half embryos and whole embryos from one of the first two blastomeres. *Anat. Anz, 10*, 623–638.

Oppenheimer, J. M. (1991). Curt Herbst's contributions to the concept of embryonic induction. In S. F. Gilbert (Ed.), *A conceptual history of modern embryology* (pp. 83–90). New York: Plenum Press.

Roux, W. (1888). Beiträge zur Entwickelungsmechanik des Embryo. Über die künstliche Hervorbringung halber Embryonen durch Zerstörung einer der beiden ersten Furchungskugeln, sowie über die Nachwickelung (Postergeneration) der Fehlenden Körperhälfte. *Virchows Arch. Anat. Physiol, 114*, 113–153.

Sander, K., & Faessler, P. E. (2001). Introducing the Spemann-Mangold organizer: experiments and insights that generated a key concept in developmental biology. *Int J Dev Biol, 45*, 1–11.

Slack, J. (2024). The organizer: What it meant, and still means, to developmental biology. *Curr. Top. Dev. Biol. 157*, 1–42.

Spemann, H. (1901). Über Korrelationen in der Entwicklung des Auges. *Verh. Anat. Ges, 15*, 61–79.

Spemann, H. (1903). Entwickelungsphysiologische Studien am Tritonei III. *Arch. f. Entw.mech. Org, 16*, 551–631.

Spemann, H. (1921). Die Erzeugung thierischer Chimären durch heteroplastische transplantation zwischen Triton cristatus und taeniatus. *Wilh Roux' Arch EntwMech Organ, 48*, 533–570.

Spemann, H. (1938). *Embryonic development and induction.* Yale University Press.
Spemann, H., & Mangold, H. (1924). Über Induktion von Embryonalanlagen durch Implantation artfremder Organisatoren. *Roux' Arch EntwMech Org, 100,* 599–638.
Stern, C. D (2024). The organizer and neural induction in birds and mammals. *Curr. Top. Dev. Biol. 157,* 43–66.
Thophkhane, S. S, & Richman, J. M (2024). Tissues and signals with true organizer properties in craniofacial developmennt. *Curr. Top. Dev. Biol. 157,* 67–82.
Von Baer, K. E. (1828). *Über Entwickelungsgeschichte der Thiere. Beobachtung und Reflexion.* Königsberg: Bornträger.

CHAPTER ONE

The organizer: What it meant, and still means, to developmental biology

Jonathan Slack[*]
Department of Life Sciences, University of Bath, Bath, United Kingdom
*Corresponding author. e-mail address: j.m.w.slack@bath.ac.uk

Contents

1. What did Spemann and his colleagues believe in 1921–22?	2
2. What was the view in the 1930s?	10
3. Evocation and individuation	12
4. Neural and mesodermal inducers	16
5. The two gradient models	17
6. Later results	19
7. Mesoderm induction	20
8. The re-emergence of experimental embryology in a "modern" guise	23
9. The second gold rush	25
9.1 Polarization of the fertilized egg	26
9.2 Mesoderm induction	28
9.3 Neuralization/dorsalization	30
9.4 Antero-posterior patterning	32
10. How we should understand the organizer today	34
Acknowledgments	35
References	35

Abstract

This article is about how the famous organizer experiment has been perceived since it was first published in 1924. The experiment involves the production of a secondary embryo under the influence of a graft of a dorsal lip from an amphibian gastrula to a host embryo. The early experiments of Spemann and his school gave rise to a view that the whole early amphibian embryo was "indifferent" in terms of determination, except for a special region called "the organizer". This was viewed mainly as an agent of neural induction, also having the ability to generate an anteroposterior body pattern. Early biochemical efforts to isolate a factor emitted by the organizer were not successful but culminated in the definition of "neuralizing (N)" and "mesodermalizing (M)" factors present in a wide variety of animal tissues. By the 1950s this view became crystallized as a "two gradient" model involving the N and M factors, which explained the anteroposterior patterning effect.

> In the 1970s, the phenomenon of mesoderm induction was characterized as a process occurring before the commencement of gastrulation. Reinvestigation of the organizer effect using lineage labels gave rise to a more precise definition of the sequence of events.
>
> Since the 1980s, modern research using the tools of molecular biology, combined with microsurgery, has explained most of the processes involved. The organizer graft should now be seen as an experiment which involves multiple interactions: dorsoventral polarization following fertilization, mesoderm induction, the dorsalizing signal responsible for neuralization and dorsoventral patterning of the mesoderm, and additional factors responsible for anteroposterior patterning.

Since its publication in 1924, the organizer experiment has been the most famous experiment in embryology. Homage is still regularly paid by citations in research papers, review articles and textbooks. Everyone knows that the organizer was an important discovery, although it has often been unclear exactly what was discovered. This article will attempt to tell the story of the organizer in terms of what people believed about it in different scientific generations. In the 1990s the organizer problem was largely solved in molecular terms, at least for amphibian embryos. This vantage point of knowing the solution provides us with useful hindsight with which to assess the beliefs of the past. In essence, I shall argue that the organizer was mysterious because it seemed to be an all-encompassing force that created the whole body plan in a single step. We can now see that the organizer experiment involves a number of distinct processes which build the body plan in successive steps: dorsalization of the fertilized egg, mesoderm induction in the blastula, the dorsalizing gradient in the gastrula, the head inducing factors, and the posteriorizing gradients. In particular, workers prior to the 1970s, were not aware of the process of mesoderm induction at the blastula stage and this made the organizer phenomenon, in those days misnamed "primary" embryonic induction, very hard to understand.

1. What did Spemann and his colleagues believe in 1921–22?

Hans Spemann (Fig. 1A) was a pioneer of microsurgery in embryos and a dominant figure in the field of experimental embryology. He started his work on early amphibian embryos at the turn of the 20th century while a junior faculty member at the University of Würzburg. This work was

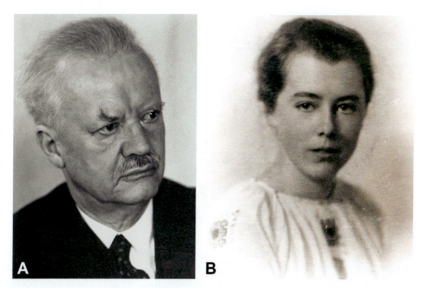

Fig. 1 (A) Hans Spemann (Nobel Prize portrait 1935). (B) Hilde Mangold in 1922 (Albert-Ludwigs Universität Freiburg, Universitätsarchiv).

continued when he moved to Rostock in 1908, Berlin in 1914, and Freiburg in 1919. There he stayed until his retirement in 1937 and death in 1941.

Prior to 1921 he had established several of the basic concepts of experimental embryology as applied to the early amphibian embryo, and the organizer experiment was planned in the context of mapping the state of determination of regions in the early amphibian gastrula. By this time, it was known from isolation of small tissue explants that the gastrula consisted of a pattern of differently specified regions. These had autonomous morphogenetic movement activity ("dynamic determination") and self-differentiation ability ("material determination"). These specified regions included those which self-differentiated into the endoderm, the mesoderm (at least its axial components) and the epidermis.

By the early 1920s the first fate mapping results from Vogt were becoming available. These showed the fate in normal development of each region of the early gastrula and so provided a basis from which interpret the various types of experiment. It was clear that in both explant isolation experiments, and transplantations of small grafts between embryos at the early gastrula stage, the later development of parts did not necessarily correspond to the normal fate. In particular, prospective neural plate (called

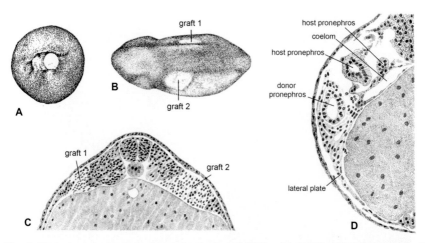

Fig. 2 Mesoderm induction shown by grafts of (Mangold, 1923). (A) A light colored graft into the marginal zone. (B) A tailbud stage embryo bearing two light colored grafts. (C) A section through the same case, showing that the grafts have incorporated into the host somites. (D) Another case, showing incorporation of the graft into the host kidney.

"medullary plate" at the time) did not self-differentiate into neural plate after isolation in simple salt solutions, although it might do so in more complex media. Instead it formed a mass of epidermis. Spemann and his assistants conducted a number of "exchange transplantations" in which small parts from different regions of two embryos were exchanged in position (Spemann, 1921). In these experiments the grafts came from light pigmented species and the host was darker pigmented (see below). The principal result was that prospective epidermis, grafted to the region fated to become neural plate, would develop as neural plate. Likewise, prospective neural plate, grafted to the region fated to become ventral epidermis, would develop as epidermis. It was also established that tissue from the animal hemisphere would become mesoderm if moved into the marginal zone (Fig. 2) (Mangold, 1923), a finding of great importance whose significance did not fully emerge for many decades.

The conclusion was that the whole animal hemisphere was "indifferent". This was a misleading term, even at the time, as the animal hemisphere was well known to be specified to become epidermis, but it conveyed the idea that it was also able to become neural plate or mesoderm if grafted to appropriate positions. In contrast to the "indifferent" animal hemisphere, Spemann had long suspected that there was something special about the dorsal lip region as

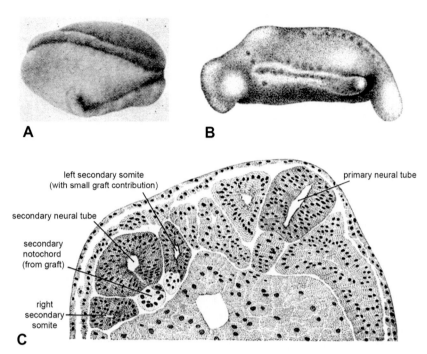

Fig. 3 The original organizer grafts. (A) Case 21 from (Spemann, 1918) which distinctly shows a secondary axis. (B) Hilde Mangold's case number 132, showing a secondary axis without head. (C) Section through case 83 showing that the entire secondary neural tube, and most of the somites, are derived from the host, and therefore have been induced.

this was such a prominent focus of convergence of cells during the tissue movements of gastrulation. Warren Lewis had earlier found that dorsal lips would self-differentiate into notochord, muscle and neural tissue when grafted to the head of a tailbud stage host (Lewis, 1907). Spemann carried out grafts of dorsal lips to other parts of the gastrula and believed that the material determination of this region was special as it not only differentiated according to its normal fate, but also recruited surrounding tissue into an embryonic axis (Fig. 3A) (Spemann, 1918). In a postscript to his later paper of 1921 dealing with the exchange transplantations (Spemann, 1921), he used the term "organizer" for the first time.

> "Such a piece of the organization center can be designated briefly as an "organizer": it creates an "organization field" of a certain orientation and extent, in the indifferent material in which it is normally located or to which it is transplanted" (translation by Viktor Hamburger).

He used the term "organization center" to refer to the region of the embryo in which the organizer is located. However, the interpretation of the dorsal lip grafts still remained uncertain because of the inability, in the experiments published in 1918, to distinguish which parts had arisen from the graft and which from the host. In the 1921 work, Spemann had started to use grafts between differently pigmented species of salamander in order to identify the origin of parts. But he had not included the dorsal lip graft in this series perhaps because at this time the fate map information was still poor and he believed that the dorsal lip region was itself prospective neural plate. The uncertainty about the contribution of graft and host following a dorsal lip graft was the starting point for the project that was to be published a few years later as the famous organizer paper (Spemann & Mangold, 1924).

Hilde Mangold, then Hilde Pröscholdt (Fig. 1B), came to the lab as a student. Her initial PhD project was to repeat a famous experiment of Abraham Tremblay in which the freshwater polyp, *Hydra*, was turned inside out. However neither she nor Spemann himself could achieve this difficult feat and so her project was altered to that of making dorsal lip grafts between gastrulae of different pigmentation. This would establish once and for all what the graft became and what effect it had on its surroundings. The work was done in the operating seasons of 1921 and 1922, during which time Hilde became Hilde Mangold when she married Spemann's assistant Otto Mangold in 1921. Tragically, Hilde Mangold died in 1924, at about the time of publication of the famous paper. She was killed as a result of a domestic accident with an alcohol stove which caused terrible burns and led to her death within a day.

At this stage it is worth mentioning some background technical facts. In these days amphibians could not be reared in the lab and there was no hormone-induced ovulation. Embryos could only be obtained in the natural breeding season which is in the spring. Usually, inseminated female newts were collected in the wild and allowed to lay their fertilized eggs in aquarium tanks in the lab. The workers would need to keep a close eye on their newts to see what was available each day of the breeding season. Of course, under these circumstances, obtaining suitably staged embryos of more than one species was especially challenging. The operations were performed in water, which is unfavorable for healing compared to the balanced salt solutions used later. Moreover, although sterile procedures were used for the experiments, there were in the 1920s no antibiotics available, so the operated embryos or explants often became infected and could easily be lost at an early stage. Under the circumstances it is remarkable that many experiments could be conducted at all. As for recording the results, photomicrography was occasionally used, but most of the illustrations in most papers of the period contain meticulous drawings of the whole specimens and of the histological sections. These were subsequently photographed for reproduction. No doubt the scientists tried their best accurately to indicate the pigmentation differences in their drawings, but there is probably an element of interpretation as well.

The main species used for the work were known as *Triton cristatus* (now *Triturus cristatus* complex, a group of 7 related species), *Triton taeniatus* (later *Triturus vulgaris*, now *Lissotriton vulgaris*), and *Triton alpestris* (now *Ichthyosaura alpestris*). The old names will be used here for clarity. Embryos of *T. cristatus*, the crested newt, are light colored, which should make them ideal as a host for dark pigmented grafts. But the embryos are large and floppy and do not develop well without their vitelline membranes. Even worse, the whole species complex carries a natural balancer chromosome meaning that only the 50% of embryos which are heterozygous for this chromosome are viable (Wielstra, 2019). Because of the shortcomings of *T. cristatus* as a host, it was usually used as a donor and its light colored grafts were visualized against the darker pigmentation of *T. taeniatus* or *T. alpestris*.

The organizer paper contains descriptions of only six cases, selected by Spemann. But the actual data set was larger (Fassler & Sander, 1996). Pröscholdt's lab notebook contains records of 259 interspecies transplantations carried out in 1921 and 1922, which yielded 28 secondary axes, 26 of which included host tissue in the secondary axis. In 1923 a further 217 grafts were done together with Otto Mangold, intended to find the spatial extent of the organizer. Of these, 17 formed secondary axes.

Of the six cases in the paper, one (case 8) is not actually an organizer graft, as it was taken far from the blastopore. The other five provide evidence for induction of the secondary neural plate from the host, and two cases (83 and 132) give clear evidence for the formation of a secondary axis in the mesoderm, with graft-derived notochord and somites derived partly from the graft and partly from the host (Fig. 3B and C). Case 132 is illustrated as a whole specimen with a clearly visible secondary axis, albeit lacking a head, and this image is often reproduced in textbooks. The work described in the paper confirmed that dorsal lip grafts did, as suspected, self-differentiate into axial mesoderm and also had a significant inductive activity resulting in the formation of an organized second body axis. With the benefit of hindsight and modern techniques, the original grafts do not

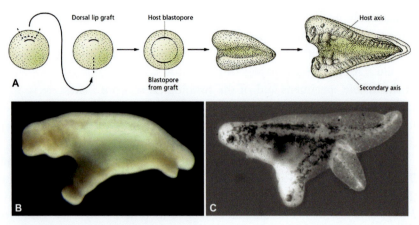

Fig. 4 The organizer graft as performed today. (A) The graft includes deep as well as superficial tissue above the dorsal lip, and is placed into the ventral marginal zone. (B,C) Organizer graft on an axolotl embryo, (B) at 5 days, (C) at 8 days.

seem very impressive. Modern organizer grafts are placed into the ventral lip and include the full thickness of the organizer tissue. These two precautions increase the chance of getting anteroposteriorly complete secondary embryos (Fig. 4).

Spemann was intrigued by the organization and direction of the secondary axes. He viewed the organizer as a source of "spreading determination" but was evidently baffled by how it might work:

"There can be no doubt but that these secondary embryonic primordia have somehow been induced by the organizer; but it cannot yet be decided in what manner this occurs, and, above all, when and in what way" (Spemann & Mangold, 1924): Translation by Viktor Hamburger.

Spemann further wrote in a review article of 1924:

"Such a secondary embryonic anlage looks as if it were built by a superior force out of material that happened to be available, without consideration of its origin or species affiliation" (Spemann, 1924): Translation by Viktor Hamburger.

It does seem as though the community were so impressed by the integrated character of the inductions and the fact that they seemed to produce a whole second body, that they ignored the existing evidence about developmental events before gastrulation, and mentally collapsed the whole of early development into a single phenomenon "the organizer". Significantly, Hamburger, writing in 1988 (Hamburger, 1988), but considering the concept "from the perspective of the late 1920s and early 1930s" felt that the grandiose title of "organizer" would not be appropriate if it referred just to one step of induction involved in building the body plan.

The consensus view at the time seems to be:

- All of the embryo except the organizer was "indifferent" up to gastrulation.
- The organizer acted to create a whole patterned body from itself and its surroundings.

In fact this view was not warranted even in the 1920s, as there was already evidence for several developmental events prior to gastrulation, which generated the regional specification of the early gastrula. But for many years, little attention was paid to these. With the benefit of hindsight, we can see things a lot more clearly. Nowadays a textbook account would start with maternal mRNA localization which creates the animal-vegetal polarity, continue with the events of cortical rotation which creates the dorsal-ventral polarity, and lead on to mesoderm induction which causes the formation of a ring-shaped zone of mesoderm around the equator of the embryo. There is a persistent blindness in the early literature about the nature of the mesoderm. The term "mesoderm" is often used just for the axial structures (notochord and somites) and the remainder (kidney, lateral plate and blood islands) is ignored. This meant that the critical result of dorsalization of the mesoderm, clearly visible in two of the cases in the paper (83 and 132), was misinterpreted. It was vaguely referred to as "assimilative induction". Although mesoderm induction (including formation of kidney and lateral plate) had already been seen by Mangold (Mangold, 1923) (Fig. 2), it was viewed as something happening as a part of the organizer's action during gastrulation. Mesoderm induction was not properly recognized as an event of development *preceding* gastrulation until the work of Nieuwkoop and Nakamura in the 1970s.

The general invisibility of the ventrolateral mesoderm, comprising the kidney, lateral plate and posterior blood islands regions, may be partly due to the fact that in newts used at the time almost all the invagination of mesoderm occurs through the dorsal lip, and the ventral lip is not at all prominent. In *Xenopus*, and other anurans, the blastopore soon becomes a complete circle and invagination clearly occurs all round it, although it proceeds much further on the dorsal side. Once molecular markers became available, it was clear that the mesoderm existed prior to gastrulation and had the form of a complete circle. In Fig. 5 are shown images of an axolotl

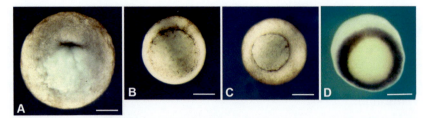

Fig. 5 Amphibian gastrulae. (A) An axolotl gastrula. The dorsal lip has the form of a pit. (B,C) *Xenopus* gastrulae. The dorsal lip appears as a concentration of pigment and rapidly becomes a complete circle. (D) In situ hybridization showing the location of *brachyury* mRNA, which is a marker of mesoderm. The extent of the mesoderm is similar all the way round. Scale bars 0.5 mm.

gastrula, in which the dorsal lip is very prominent, two stages of *Xenopus* gastrulation, in which the lateroventral invagination is clearer, and an in situ hybridization of a *Xenopus* early gastrula showing expression of the mesodermal transcription factor gene *brachyury*. The ring of mesoderm is symmetrical at this stage although it later becomes more concentrated on the dorsal side as a result of the tissue movements of gastrulation.

2. What was the view in the 1930s?

The 1930s were the years of the famous "gold rush" for the chemical basis of the organizer. This started in 1932 when a short paper appeared in *Naturwissenschaften* with sections by Bautzmann, Holtfreter, Spemann and Mangold (Bautzmann et al., 1932). Each had, in somewhat different ways, shown that *killed* organizer tissue possessed inductive activity. The tissues were implanted into gastrulae by making a hole in the animal cap and pushing the implant into the blastocoel. This method ("Einsteckung"), introduced into the lab in the 1920s, was a quick and simple method of bringing an implant into contact with ectoderm, easier than the organizer graft which involved implantation into the surface of the gastrula (ideally into the ventral marginal zone). This sensational result strongly suggested that the organizer activity had a chemical basis, and was probably some sort of diffusible substance.

Shortly after, Holtfreter conducted a mammoth series of experiments using an even simpler test method: called "Umhüllung" (Holtfreter, 1934a, 1934b). This involved wrapping the test tissue in isolated ectoderm followed by culture in a buffered salt solution and later became known as the "sandwich" procedure. There were two principal results. First, other regions of the embryo aside from the organizer could become active when killed, indicating the presence of "masked" inducers in these areas. Secondly, a wide variety of living or killed animal tissues from many sources were also active, indicating that the inductive activity was very widespread in nature. Several groups from Germany and Britain then attempted to purify the activity (Fischer & Wehmeier, 1933; Waddington et al., 1934). Over the next few years many classes of substance were claimed to be active, but the work was eventually unsuccessful. We now know that the biochemical techniques of the time were hopelessly inadequate for this task. The purification of growth factors from tissues requires at least a 10^6 fold purification and this needs multiple affinity and HPLC columns together with a simple bioassay for activity. It also needs a lot of starting material, not just a few dissected embryos.

By 1935, pure substances were being tested and some of them were shown to have activity. They included polycyclic hydrocarbons, of great contemporary interest as they were among the recently discovered chemical carcinogens. Eventually tissues treated with methylene blue, a synthetic dyestuff which does not exist in nature, were shown to be active. The lack of specificity in the signal caused a serious rethinking of the problem. It was generally concluded that the test tissue, the gastrula ectoderm, was very labile in its determination, and a slight disturbance from many factors could bring about neural inductions. This conclusion, that the signal was too non-specific to be isolated by biochemical purification, remained widely believed until the 1980s.

Throughout the gold rush it had tended to be forgotten that the organizer was the source of a complex field of organization, and the attention had focused just on the induction of neural plate. Needham, who had been very active in this area, later discussed the matter at great length in his book *Biochemistry and Morphogenesis* (Needham, 1942). Along with his co-worker Waddington, he pointed out that the inductions by killed tissue and by pure substances tended to be simple neural inductions, while those from live dorsal lip were much more complex. They called the neural inducing activity the "evocator", underlining that the capacity for neural differentiation was present in the responding tissue and was simply activated, or "evoked", by what was probably a rather unspecific stimulus. The contrast is indeed very obvious between the inductions shown in Fig. 6 from a killed tissue, a pure substance, and a living dorsal lip graft.

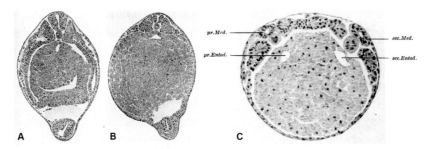

Fig. 6 Evocation and individuation. (A) Simple neural tube induced by an implant of boiled oocyte nucleus. (B) Simple neural tube induced by an implant of the carcinogen benzpyrene (Needham, 1942) (C) Complete secondary axis induced by an organizer graft into the marginal zone. Case 164 of Mayer (Mayer, 1935).

The much vaunted non-specificity of neural induction had actually been exaggerated and the differences between species had been glossed over. We now know that neural induction is provoked by very few factors in *Xenopus* ectoderm, and probably that of other anuran species, unless the tissue is first dissociated in calcium-free medium. However, urodeles such as the axolotl are prone to occasional spontaneous neuralization, suggesting that in them the situation is much more finely balanced (Fig. 7C). In this regard, the newt species that were used in the 1930s were probably more like the axolotl than like *Xenopus*. Also, we now know that wounding itself activates signal transduction pathways which are also activated by inducing factors (Christen & Slack, 1999) so the extent and type of wounding involved in the procedure is also likely to make a difference.

3. Evocation and individuation

Needham and Waddington were both heavily involved in the gold rush. In addition, Waddington had discovered a comparable phenomenon in birds indicating that the organizer was something that was not confined to amphibian embryos. He stressed the importance of the response to an inductive signal, which he called "competence" (Waddington & Gray, 1932). Both authors distinguished clearly between "evocation", as displayed in the neural inductions, and what they called "individuation", which was the conferring of an organized shape and regional structure to the secondary embryo. The complex process of individuation clearly had a lot to do with competence.

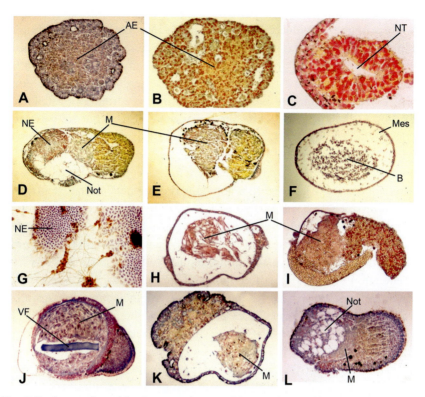

Fig. 7 Explant and combination experiments. (A) Animal cap of *Xenopus*, cultured for three days. It forms a mass of epidermis. (B) Animal cap of axolotl, also forms epidermis. (C) Spontaneous formation of a neural structure in another axolotl animal cap culture. (D) Dorsal marginal zone explant from *Xenopus*. It has formed notochord and a muscle mass. (E) Intermediate marginal zone explant. It has formed a large muscle mass. (F) Ventral marginal zone explant. It has formed mesenchyme and blood cells. (G) A culture of disaggregated *Xenopus* animal cap cells on a laminin-fibronectin substrate. Neuroepithelium has formed, together with some neurons bearing axons staining for an antibody to neurofilament protein (brown). (H) Result of a transfilter mesoderm induction. A *Xenopus* animal cap explant has formed a muscle mass after exposure to a diffusible signal from a vegetal explant. (I) Dorsalization by the organizer. A ventral marginal explant from *Xenopus* was cultured in combination with a dorsal marginal explant from an axolotl (large nuclei, right side). The formerly ventral *Xenopus* tissue has formed a large muscle mass. (J) Mesoderm induction in an animal cap from an implant of Tiedemann's vegetalizing factor. (K) Mesoderm induction of an animal cap by recombinant eFGF. (L) Mesoderm induction by activin, note the formation of notochord. Labels: AE atypical (i.e. non-stratified) epidermis; NT neural tube; NE neuroepithelium; M muscle; Not notochord; Mes mesenchyme; B blood; VF vegetalizing factor.

How might individuation work? In his 1936 book, issued as an English edition in 1938 (Spemann, 1938), Spemann discussed the various theories that might bear on the problem. He started with Driesch who, in the late nineteenth century, had shown that separated sea urchin blastomeres could form complete miniature embryos. He dismissed Driesch's "Entelechy" concept as vitalistic and lacking in explanatory power, but he was more sympathetic to Driesch's conception of a "harmonious-equipotential" system, also employed by Harrison in the USA in relation to limb development. A harmonious equipotential system is a region of cells which have a similar potency. If the region is augmented or reduced in size, it adapts to the disturbance and still produces a normal set of structures, maybe of augmented or reduced size. Exactly how such regulatory behavior arose was left unclear. The next relevant conception was the "embryonic field", advanced by Weiss to explain similar regulative phenomena in newt limb regeneration (Weiss, 1925). This involved an analogy with magnetic fields, which conserve polarity and some pattern on fragmentation, but again had limited explanatory power. More promising was the gradient theory. This had originated from Boveri in the 1890s in relation to centrifugation experiments on the early sea urchin embryo, and the *Ascaris* (nematode) embryo. The principal exponent of gradients was Child in the USA, who produced a most influential theory based on gradients of metabolic activity (Child, 1928). These could actually be observed, through the change in color of redox dyes added to sea urchin embryos, and also by the disintegration gradients of planaria observed following treatment with metabolic inhibitors (Slack, 1987). The gradient theory and the embryonic field concept were combined into a single theory by Huxley and de Beer (Huxley & de Beer, 1934), applied mainly to the regeneration of hydroids. Unfortunately, the emphasis on metabolism as the basis for the gradient led Spemann and others to reject this as an explanation for the organizer phenomenon. Many experiments had been carried out involving unilateral heating of early embryos, and thus the perturbation of metabolic rate in particular parts of the embryo. These could change the relative rates of cell division and other events but did not, in general, affect the pattern developed by the embryo.

Not until 1937 was a gradient theory advanced based on the idea of gradients of bioactive substances, with threshold responses. This was originated by Dalcq and Pasteels (1937). They proposed that a gradient of a yolk-associated substance, V, extended from the vegetal to the animal pole. And another gradient of a cortical substance, C, extended from the dorsal

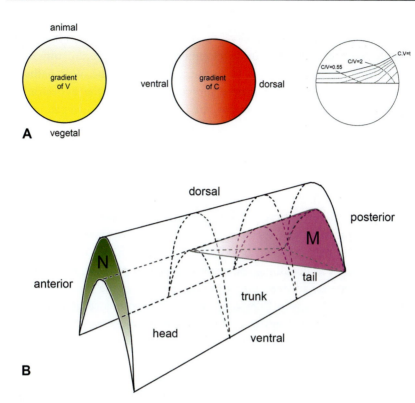

Fig. 8 Two-gradient models. (A) Model of Dalcq and Pasteels (1937). Threshold responses to C.V and C/V can generate an approximation to the fate map of the mesoderm in the cleavage stage embryo. (B) The model of Toivonen and Saxén (Toivonen, 1953). The neural plate is patterned by a dorsal-to-ventral gradient of an N factor, and a posterior-to-anterior gradient of an M factor.

to the ventral side. It had been known for some time that following fertilization a pigment shift occurred leading, in some species, to the appearance of a "gray crescent" on the prospective dorsal side, and this was considered to be the high point of the C gradient. The value of the product CV determined the appearance of the blastopore lip, and the value of C/V determined the dorsoventral character of the mesoderm (Fig. 8A). Their model explicitly required the existence of threshold responses and it was proposed that invagination behavior required a value of C.V > t, with the time of invagination being earlier the higher the value. The ratio C/V was suggested to encode notochord at C/V > 2, and somites when C/V > 0.55. Below this value the mesoderm would become lateral plate and blood islands. The relevant thresholds provide a good fit to the prospective regions for these tissues on the fate map.

This model is remarkably close to our modern understanding of the pre-gastrula phase of amphibian development, with V approximating to the mesoderm inducing signal and C to the activation of the Wnt pathway. Unlike the conception of a simple substance acting as the evocator and inducing neural plate from ectoderm, the Dalcq-Pasteels model can explain the formation of a complex pattern from a plausibly simple starting position. Unfortunately, the model does not seem to have been seriously applied to the organizer phenomenon. Perhaps the thinkers of the time considered that it really was just a model for the immediately post-fertilization events and had no particular relevance to the organizer.

Spemann and others were impressed by the "wholeness" of the secondary embryos produced by the organizer graft. But they were by no means always "whole". Often they consisted of isolated parts of the head, the trunk or the tail, and, from the beginning, attention was given to how the organizer managed to generate an anteroposterior body pattern. In the early work it was established that early gastrula dorsal lips would tend to induce head structures while late gastrula dorsal lips tended to induce tail structures. Spemann also highlighted host effects, especially the ability of late lips transplanted to an anteroventral position to generate heads (Spemann, 1931). The ability of induced neural plate to induce more neural tissue (homeogenetic induction) had earlier been discovered (Mangold & Spemann, 1927), and when pieces of neural plate from neurulae, or of the underlying archenteron roof mesoderm, were inserted into early gastrulae, they would induce neural structures with anteroposterior specificity (Bautzmann, 1929; Mangold, 1929, 1933). In addition an experiment of Holtfreter (Holtfreter, 1933) showed that when gastrula ectoderm pieces were placed on the endo-mesodermal part of an exogastrula (an embryo reared in isotonic salt in which the endo-mesoderm separates from the ectoderm), the inductions also had an appropriate anteroposterior character.

So, in addition to its other remarkable properties, the organizer could somehow generate a body pattern in the anteroposterior axis.

4. Neural and mesodermal inducers

The next chapter in this story starts at the end of the 1930s with a discovery by Chuang (Chuang, 1939) that animal tissues showed two distinct types of inducing activity: neural and mesodermal. This was confirmed by voluminous work of Toivonen (e.g. his later publication

(Toivonen, 1953)). The two inducing activities had different representations among the many tissues that were tested, and the mesodermal factor was destroyed rapidly by boiling whereas the neural factor was relatively stable. 1939 also, of course, marked the beginning of the Second World War and the resulting disruption largely terminated the activity of the German school which had led the way up to this time.

The researches of the 1940s and 1950s are summarized in a monograph by Saxén and Toivonen entitled "Primary Embryonic Induction" (Saxen & Toivonen, 1962). The title of this monograph once again indicates the relative obscurity of the pre-gastulation events of amphibian development and indicates that the organizer phenomenon is the "primary" event. It also fatally confuses what we now call "mesoderm induction" with "posteriorization", or what was at the time called "spinocaudal induction". This was based partly on the fact that the most posterior part of the neural plate is actually mesodermal in fate, and eventually also in determination, forming as it does the mesodermal parts of the tail. Because two classes of inductor were known, N and M factors, it seemed reasonable to conclude that the balance between them, or rather their endogenous counterparts, was responsible for the anteroposterior pattern of the neural plate.

There was little progress in the identification of the putative N factor. Confusion persisted about its chemical nature, many classes of substance were active and many inhibition experiments gave contradictory results. By contrast more uniform results were obtained for the M factor. This was clearly a protein. It was soluble in water, it was somewhat heat labile, it was destroyed by proteases and chemical treatments that destroy protein side groups. Purification attempts were made by three groups, that of Saxén and Toivonen in Helsinki, Yamada's in Nagoya, and Mangold's group at the Heiligenberg Institute. Because of his Nazi sympathies, Mangold had been removed from his post in Freiburg after the war and continued his work in a small privately funded institute at Heligenberg, near Lake Konstanz. In 1953 he was joined by the biochemist Heinz Tiedemann who, much later, and then working in Berlin, achieved complete purification of the M factor extracted from mid-stage chicken embryos (Plessow et al., 1990).

5. The two gradient models

The model of Dalcq and Pasteels was not the only two gradient model. A rather prescient study on patterning within the mesoderm was produced in Germany by Tuneo Yamada (Yamada, 1937, 1940). The work, which was continued later in Japan, was conducted on slit blastopore and neurula stages of *Triturus* spp. In essence, when pieces of the mesodermal layer were grafted to a more ventral position, they differentiated in a more ventral manner than their position of origin. For example, prospective somite would form kidney, and prospective kidney would form blood islands. If accompanied by a piece of notochord, the mesoderm would become dorsalized, for example prospective blood islands would become kidney and prospective kidney would become blocks of muscle. In general, the mesoderm was labile until the beginning of neural plate formation and then became determined. Similar results were obtained with explant and combination experiments with the tissues being wrapped in an ectodermal sandwich. Yamada later proposed a "double potential" theory with a variable called Pcc (cephalocaudal) representing morphogenetic movement and Pdv (dorsoventral) representing the intensity of metabolism (Yamada, 1950). These gradients would account for the regional pattern of induction by the invaginated mesoderm and the character of the resulting neural structures.

This theory was rather ahead of its time in that it closely resembled the later two factor models of Nieuwkoop and Saxén/Toivonen. But it explicitly related to the neurula stage of development, not the gastrula. It only became clear much later that the dorsoventral gradient with the presumptive notochord as the signaling center, so clearly shown in these experiments, is actually a key attribute of the organizer itself.

Work by several groups had started to indicate a posterior dominance in anteroposterior specification but the most influential was by Nieuwkoop in a series of three papers published together in 1952 (Nieuwkoop, 1952a, 1952b, 1952c). Working with various urodele embryos, folds of gastrula ectoderm were implanted vertically into different anteroposterior levels of the neural plate of the host. The inductive signal travels some way up the fold and produces an ordered set of neural structures, with the most posterior being appropriate to the level of implantation, and progressively more anterior structures appearing more distally. This looks very much like a posterior to anterior gradient, but Nieuwkoop preferred to interpret the results in terms of two signals. This was because the anterior inductions were larger, although contained fewer structures, and the posterior inductions were often distally incomplete, as well as being smaller. He proposed that there was an "activation" signal that induced neural plate, and in the absence of further signals this would become forebrain. There was also a "transformation" signal emanating from the posterior region, which, depending on its strength, made the inductions progressively more posterior.

After much examination of the activity of different animal tissues, Toivonen had settled on guinea pig liver as a good source of N activity, and bone marrow as a good source of M. A very influential experiment reported in 1955 involved implanting pellets of both tissues, killed by alcohol treatment, into early gastrulae, which resulted in anteroposteriorly complete inductions. A two gradient model emerged from this which was rather different from that of Dalcq and Pasteels, and achieved a much greater visibility (Toivonen & Saxen, 1955). The model was presented in terms of a diagrammatic neurula with the N factor being graded from dorsal to ventral, and the M factor from posterior to anterior (Fig. 8B). Once again, the M factor is considered as a spinocaudal or posteriorizing inducer, not as an agent inducing the mesoderm in the blastula.

Embryology books of the 1950s and 1960s devote plenty of space to the organizer. Waddington, who had been actively involved in studying the organizer in birds, and in the first biochemical gold rush, presents the issue largely as it was perceived in the 1930s, except that he discusses his own concept of competence at much greater length (Waddington, 1956). Balinsky (Balinsky, 1960) (later editions 1965 and 1970) treats the organizer as an agent of neural induction, with anteroposterior regional specificity. The dorsoventral gradient of Yamada is mentioned but not clearly distinguished from the caudalizing signal of Toivonen. By this date the model of Toivonen and Saxén has captured the imagination and is being advanced as the most likely mechanism to explain what is happening.

6. Later results

Niu and Twitty (1953) were interested in the issue of whether induction really required cell contact, something that had been generally accepted up until then. Using very tiny explants of epidermis which were allowed to spread on glass coverslips, they showed that medium previously conditioned by culture of mesoderm was capable of inducing neurons and pigment cells. Later, in 1961 Saxén performed transfilter experiments in which the inducer and responding ectoderm were separated on opposite sides of a thin but porous membrane (Saxén, 1961). He obtained neural but not mesodermal inductions. This was consistent with the neural inducing factor being a diffusible substance. Later still, these experiments were refined with the use of Nuclepore membranes, which have a better defined pore size, and it was found that the mesodermal signal could also pass through a small pore membrane (Grunz & Tacke, 1986) (Fig. 7H).

A series of studies by Barth and Barth, of which the most highly cited is (Barth & Barth, 1959), investigated the self-differentiation of very small ectoderm explants from *Rana pipiens*. These were cultured in an improved version of the Niu-Twitty procedure in which the medium contained some inert protein (globulin) to improve cell survival. The explants were partially disaggregated in Ca-free medium including EDTA, and the outer impermeable layer was removed. Small numbers of cells were cultured on glass coverslips. Although explants of *Rana* ectoderm do not show auto-neuralization, these cultures produced neurons and pigment cells. When treated with lithium, additional cell types were seen. This study increased the general feeling that inducing factors were non-specific, and raised the possibility that even simple ions might have inductive effects.

Another interesting result of the post-War period came from the cortical grafting experiments of Curtis (1960, 1962). These experiments were carried out to investigate the "cortical field" concept of Dalcq and Pasteels (1937). Small explants of cell cortex from the dorsal side of the late stage fertilized egg were grafted to the ventral side of recipients, and, with high frequency, provoked the formation of secondary embryos. These experiments did make some impact at the time as they seemed to push back the time of organizer formation and to make it seem like a cytoplasmic determinant, similar to the pole plasm of *Drosophila*, or the polar lobe of molluscs. However, the experiments were discredited by later studies on the cortical rotation ((Gerhart et al., 1981) and see below), which maintained that Curtis' positive results were artifacts of tipping the demembranated embryos ventral side up, such that a cortical rotation was initiated in opposition to that provoked by sperm entry. Despite this, similar results to Curtis' have been obtained more recently (Kageura, 1997) and it remains possible that Curtis was correct.

7. Mesoderm induction

The really critical missing part of the puzzle was filled in the 1960s by work in Holland and Japan. Nieuwkoop in Holland used mostly the axolotl and Nakamura in Japan used *Triturus* and *Xenopus,* but the results

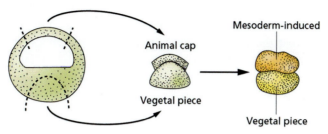

Fig. 9 Mesoderm induction. The process is most clearly demonstrated in "Nieuwkoop combinations" of animal and vegetal tissue from the blastula.

are similar. Nieuwkoop's experiments started with the division of the blastula into 4 equal rings (1, 2, 3, 4) from the animal to the vegetal pole (Nieuwkoop, 1969). A deformed, but approximately normally patterned, embryo containing a mesodermal axis arose only from ring 3 (sub-equatorial). However if rings 1 + 2 + 4 (animal hemisphere + vegetal mass) were recombined, a similar structure arose. Later it became clear that mesodermal structures were formed if a moderately small animal and vegetal explant were recombined (Fig. 9). Using interspecies grafts, or grafts labeled with tritiated thymidine, Nieuwkoop showed that the mesoderm arose from the animal hemisphere component (Nieuwkoop & Ubbels, 1972). Furthermore, the dorsoventral specificity of the induced structures was a property of the vegetal component, so axial mesoderm was only induced by the dorsovegetal region, and lateral plate mesoderm was induced by the remainder of the vegetal mass (Boterenbrood & Nieuwkoop, 1973). The dorsovegetal region later became known as the "Nieuwkoop center". The results of the Japanese workers were similar (Nakamura et al., 1971). They also showed that self-differentiation of the prospective marginal zone does not commence unless it is isolated from stage 7 or later (early blastula) (Nakamure & Matsuzawa, 1967). This suggested that it arose through an inductive interaction after stage 7. Using recombinants with components of different stages they concluded that the vegetal signal commenced about stage 8 (mid-blastula) while competence of the animal hemisphere was lost by about stage 11 (mid-gastrula). These timings do vary with species, with slower developing species showing inductive events occurring somewhat later relative to morphological stage.

Although usually called "mesoderm induction" both Dutch and Japanese workers noted that endodermal structures can be induced by strong signals and so the term "mesendoderm induction" is perhaps more appropriate. There was some terminological dispute with the Dutch group maintaining that the mesoderm was "induced by the endoderm from the ectoderm", while the Japanese viewed the events as "regulation of an animal-vegetal gradient", but the phenomena described are clearly the same. It should perhaps be noted that Waddington had previously shown an induction of the primitive streak by the endoderm in the chick embryo, although he viewed this more as an effect on cell movements ("dynamic determination") rather than a change of material determination (Waddington, 1933).

By the 1970s, a number of developmental biology, rather than strictly embryology, textbooks were appearing (Berrill & Karp, 1976; Browder, 1980; Ede, 1978; Grant, 1978; Ham & Veomett, 1980), and the contemporary beliefs about the nature of the organizer may be assessed by looking at them. The organizer experiment still features prominently in all of them, but only as an agent of neural induction. Nieuwkoop's work on mesoderm induction had been published by then, but does not figure in these textbooks, except for that of Grant, where it is not connected to the organizer. The books do all communicate the fact that the organizer somehow imparts anteroposterior pattern to the neural plate, and in most cases the model of Toivonen and Saxén is presented as a possible explanation. These books tend to be skeptical about the significance of the numerous heterologous inducers, and to stress the presumed nonspecificity of neural induction.

The field was summarized yet again in a volume edited by Nakamura and Toivonen (1978) to mark the 50th anniversary of the organizer experiment. This involved as principal contributors the heads of the Japanese, Finnish and German labs that were still active and covered the familiar themes: the discovery of the organizer, the gold rush for the inducing factors, the two gradient models of neural induction, and also comparable phenomena in other vertebrates. But this book also included a chapter on "Epigenetic formation of the organizer" because by this time the work by Nieuwkoop and Nakamura's labs had clearly documented the process of mesoderm induction. Although the volume is devoid of "modern" thinking in the sense used below, this critical part of the puzzle was now in place. But they still did not seem to shift the consensus represented by the textbooks. Opinion seemed firmly settled on the idea that the organizer graft was an agent of neural induction, that it imparted anteroposterior polarity, and that the two gradient theory of Toivonen and Saxén provided the best explanation.

During the long postwar period, from about 1950 to 1980, it is notable that the community of workers had remained somewhat static. The Japanese, Finnish, Dutch and German groups continued to work within the existing limits of speculation. Their results were published in what would now be called low impact journals, and attracted little attention from outside. The "outsiders": Niu/Twitty, Barth/Barth and Curtis had added a little to the story but did not make a big impact on it. The principal labs cited each others' work extensively and essentially the same authors were citing each other in 1980 as in 1950.

8. The re-emergence of experimental embryology in a "modern" guise

By the 1970s a new influence emerged of theoretical models that aimed to provide a plausible explanation in genetic and biochemical terms for how developmental processes worked. Foremost was the "positional information" world-view of Wolpert (Wolpert, 1969) but there were many others.

The first "modern" work on the organizer was performed by Cooke (1972a, 1972b, 1972c, 1973a, 1973b). He viewed the organizer as the apex of a field of positional information and investigated the effects of competition between two organizers, of excision of the organizer, and of the inhibition of cell division. Notably, he used grafts which included the full thickness of the region around the dorsal lip, and he implanted them to specific positions around the marginal zone, not into the blastocoel or the animal hemisphere as many previous workers had done. He also used *Xenopus*, which by then had become a domesticated laboratory animal from which eggs could be obtained all year round by hormonal stimulation. Cooke's diagrams explicitly assumed a dorsal-to-ventral gradient of positional information emanating from the organizer. Although Cooke certainly belongs to the modern era, he may have neglected too much of what went before. He does reference Nieuwkoop's work on mesoderm induction but does not discuss its significance. His treatment of the organizer is also somewhat detached from the task of understanding the stepwise process by which the body plan is constructed.

In the late 1970s injectable intracellular lineage labels became available. These were large molecules unable to pass between cell through gap junctions, which were non-toxic and which could be fixed for

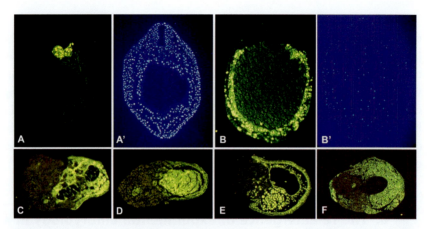

Fig. 10 Use of fluorescein-dextran amine (FDA) as a lineage label (Dale & Slack, 1987b). (A,A′) Fate mapping the dorsal marginal zone in an axolotl. (B,B′) Fate mapping the ventral marginal zone in an axolotl. (C–F) inductive processes in *Xenopus*. In each case the responding tissue was labeled with FDA. (C) Dorsal mesoderm induction. (D) Intermediate mesoderm induction. (E) Ventral mesoderm induction. (F) Dorsalization of ventral marginal zone by dorsal marginal zone. A large muscle mass has formed instead of blood islands.

histological analysis. The first was the enzyme horseradish peroxidase, previously used for neuronal tracing. This was introduced for *Xenopus* embryos by Marcus Jacobson (Jacobson & Hirose, 1978), but unfortunately he did not believe in the existence of embryonic induction at all. Soon after, Gimlich and Cooke introduced a superior label, fluorescein dextran amine (FDA), and showed that induction did, indeed, occur in *Xenopus* organizer grafts (Gimlich & Cooke, 1983).

My own lab then used the new lineage labels in an extensive study of early *Xenopus* development (Dale & Slack, 1987a; Dale et al., 1985; Smith & Slack, 1983) (Figs. 10 and 11). In these days in the broad cell biology community, as opposed to the rather closed group of traditional workers discussed above, there was much skepticism about the very existence of embryonic induction, so repeating the key experiments in this new way was of some value in correcting the misperception. We produced a new fate map, studies on mesoderm induction and regionalization, and a study on the organizer. The data were used to construct the "three signal model" (Fig. 12A). According to this, mesoderm induction in the blastula came in two qualities, a small region (the aforementioned "Nieuwkoop center") which could induce organizer tissue, and a large region which induced

The organizer and its meaning

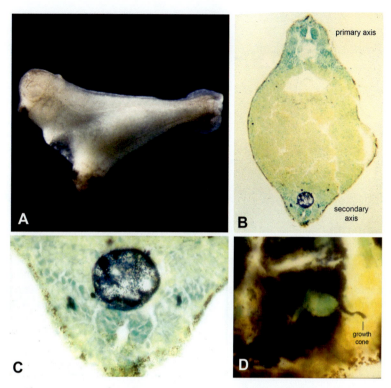

Fig. 11 The organizer graft in *Xenopus*, using HRP as a lineage label (Smith & Slack, 1983). (A) An organizer graft showing a complete secondary axis. (B,C) Transverse section of a case with HRP labeled donor. This has populated the notochord of the secondary axis, but the neural tube and almost all the somites are induced form the host. (D) A neural tube induced from an HRP-labeled host. An axon with growth cone is visible.

ventrolateral mesoderm. During gastrulation, the graded dorsalizing signal from the organizer divided up the mesoderm into territories committed to form somites, kidney, lateral plate and blood islands. At this stage it was unclear whether neural induction resulted from the same dorsalizing signal acting on the ectoderm, or from a different signal altogether. This model did catch on to some extent and served as a basis for evaluating possible molecular players in the mechanism.

9. The second gold rush

The problems were finally largely solved by the identification of the various signals. This was done partly by protein purification, partly by

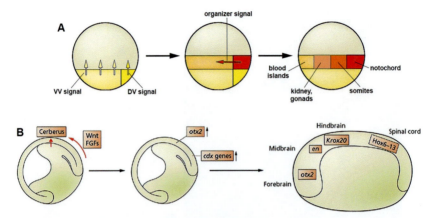

Fig. 12 (A) The three signal model for early regional specification. Following the cortical rotation, the DV region (the Nieuwkoop center) emits a DV signal inducing the organizer. The remainder of the vegetal tissue induces a ring of ventral type mesoderm. During gastrulation, a dorsalizing signal from the organizer divides the mesoderm into zones forming somite, lateral plate, and blood islands. (B) Anteroposterior patterning. Head-inducing factors such as Cerberus induce expression of forebrain genes, such as *otx2*. Gradients of Wnt and FGF induce expression of CDX and HOX genes to specify the trunk and tail.

inspired guesswork and the testing of candidates, and partly by the completely new method of expression cloning from cDNA libraries. By this time molecular biology was being seriously employed in developmental biology and a method had been invented to overexpress single gene products in *Xenopus* embryos. Synthetic mRNA was prepared in vitro (Krieg & Melton, 1984) and injected into fertilized eggs or individual blastomeres of cleavage stages. This provided new methods for testing candidates by overexpressing candidate mRNAs in animal caps, or in UV embryos (see below) or on the ventral side of normal embryos.

For clarity the experiments will be described in order of developmental mechanisms not in historical chronology.

9.1 Polarization of the fertilized egg

The first process is the dorsoventral polarization of the egg that occurs after fertilization. Investigation of this has a long history and the data available at the time informed the two gradient model of Dalcq and Pasteels (1937). It was thoroughly reinvestigated by Gerhart et al. (1981) who concluded that the critical event was a rotation of egg cortex relative to interior cytoplasm by about 30° toward the point of sperm entry. This applies only to *Xenopus*, and probably other anura. Urodeles often show polyspermy and so although there may be a cortical rotation, the role of the sperm is different. Gerhart's study tended to downplay the existence of any dorsal determinants, and discredited the experiment of Curtis, as similar double embryos could be obtained simply by demembranating and inverting fertilized eggs. Soon after it was found that an aligned mass of microtubules was an essential component of the cortical rotation. The cortical rotation could be blocked by agents that disrupted microtubules, particularly UV irradiation of the vegetal hemisphere shortly after fertilization (Scharf & Gerhart, 1983). The most extreme UV embryos were radially symmetrical and extreme ventral in character, i.e. all the mesoderm consisted of mesothelium and blood islands, indicating that the formation of the Nieuwkoop center, and the organizer, had been prevented.

The reverse phenotype, of an extreme dorsoanterior structure, effectively an embryo with organizer tissue all around the marginal zone, could be produced by treatment of early stages with lithium (Kao et al., 1986). Lithium has a long history in the treatment of many types of embryos. In the case of amphibian embryos it had mostly been applied to gastrula stages, at which it produces a ventro-posteriorizing effect (Lehmann, 1937), although the hyperdorsalized structures had been seen before (Bäckström, 1954). It was generally considered that lithium had some sort of weak mesodermalizing or vegetalizing effect (Masui, 1961). Not until 1989 was it clarified that early lithium treatment had a dorsoanteriorizing effect, while late (after the mid blastula stage) treatment had a ventroposteriorizing effect (Yamaguchi & Shinagawa, 1989). This is due to the presence of different repressors of Wnt targets which are active at the two stages: Tcf-3 early and Lef-1 late (Roël et al., 2002).

Soon after the cortical rotation mechanism had been worked out, it was shown that components of the Wnt signaling pathway could induce second axes when injected into UV embryos or the ventral side of normal embryos (Smith & Harland, 1991; Sokol et al., 1991). The first of these studies used the technique of expression cloning. cDNA libraries were constructed from lithium-treated embryos and fractionated to find an active component. The canonical Wnt pathway involves an inhibition of the enzyme glycogen synthase kinase 3, which inhibits phosphorylation of β-catenin. Unphosphorylated β-catenin is free to enter the nuclei and, by derepressing the action of other transcription factors (Tcfs and Lefs), activate gene expression. It was found that when oocytes were depleted of β-catenin mRNA by antisense treatment, then matured to eggs and fertilized, that the resulting embryos had no dorsal axis. Activity could be rescued with excess β-catenin RNA (Heasman et al., 1994). Transplantation of cytoplasm or cortical material then confirmed that a dorsal determinant was indeed shunted from the vegetal pole region to the dorsal side during the cortical rotation (Kageura, 1997). The actual dorsal determinant consists of multiple Wnt pathway components moved dorsally by the motor protein kinesin attached to the microtubule array (Larabell et al., 1997). One recently identified component is the Wnt agonist *huluwa* (Yan et al., 2018).

Considering the long history of lithium treatment of embryos, and the important role lithium had played in discovering the mechanism of early dorsoventral determination, it was ironically not until 1996 that the relevant biochemical action of lithium was at last identified (Klein & Melton, 1996), as being an inhibition of glycogen synthase kinase 3, and thus stimulation of the Wnt-β-catenin pathway. Previously, attention had centered on its inhibition of inositol monophosphatase, and before that, on competition with sodium ions. In this instance, developmental biology had solved a problem for biochemistry and pharmacology, rather than the reverse.

In summary, the investigations of events following fertilization showed that the Wnt pathway is activated in a dorsal sector spanning the Nieuwkoop center and the future organizer.

9.2 Mesoderm induction

The next developmental event is mes-endodermal induction. A candidate for the biochemical basis for this was of course the vegetalizing factor of Tiedemann. But, after 30 years of purification activity of this factor from a heterologous source, the mid-stage chick embryos, this was no longer being taken very seriously by those developmental biologists who knew about it. Exceptions were myself and Jim Smith, who did take it seriously after I requested a sample of the factor and found that it did indeed have strong mesoderm inducing activity in the sandwich assay (Fig. 7J). This stimulated a few of the "modernists" to think more seriously about mesoderm inducing factors and led to a new type of assay. Instead of implanting pellets into embryos or ectoderm sandwiches, we started to use an end point dilution assay whereby animal caps were treated in solution with serial dilutions of an active fraction and the highest dilution producing an induction was considered to be the titer. Sometimes molecular markers were used to make the process look more modern, but the workers soon found that they could just as easily score the explants by looking at them down the dissecting microscope. An important breakthrough was made by Smith, who found that a *Xenopus* cell line, XTC, secreted a potent factor that gave organizer-style inductions (Smith, 1987) (Fig. 7L). Another was made by myself, who tested a panel of candidate growth factors and found that fibroblast growth factors (FGFs) had strong activity as ventral mesoderm inducers (Slack et al., 1987) (Fig. 7K). Purification of the XTC factor showed that it was activin, formerly known as a hormone which increased FSH secretion by the pituitary gland (Smith et al., 1990). Almost simultaneously, activin was shown to be active in a screen of TGFβ family factors (Asashima et al., 1990), and a little later the old vegetalizing factor of Tiedemann was also finally identified as activin (Tiedemann et al., 1992).

There followed a mad rush of activity by many labs: testing other known factors in all possible combinations, and cloning activin-like and FGF-like genes from *Xenopus*. Initially, from the observed biological activities, it was suspected that activin was the DV signal from the Nieuwkoop center, and FGF was the VV signal from the remainder of the vegetal mass. Overexpression of dominant negative inhibitors inhibited mesoderm formation and this suggested that both classes of factor were indeed needed for the process in vivo (Amaya et al., 1991; Hemmati Brivanlou & Melton, 1992).

But things were no so simple. It turned out that there is not much actual activin in the early embryo, but various similar proteins, called "nodal-relateds" do have a similar biological activity to activin and are expressed in the vegetal hemisphere. These, collectively, seem to make up the mesoderm inducing signal, which is relatively weak, producing a ventral-type induction. In the organizer region they synergize with the activated Wnt pathway to give a stronger, i.e. organizer-inducing, effect. The best evidence that nodal-related factors are the endogenous mesoderm inducing factors came from inhibition experiments using "Cerberus short", a protein fragment which inhibits nodal-related factors but not activin or various other candidates (Agius et al., 2000). This inhibits organizer formation in whole embryos and mesoderm induction in animal-vegetal combinations. The FGFs turned out not to be the VV signal but they are nonetheless required for mesoderm induction, partly by increasing the competence of the animal hemisphere and partly by being involved in a positive feedback loop with the mesodermal transcription factor Brachyury.

9.3 Neuralization/dorsalization

The next event, occurring during gastrulation, is the emission of a dorsalizing signal from the organizer with the consequent induction of the neural plate from the ectoderm, and the induction of a series of territories in the mesoderm which will form somites, kidney, lateral plate and blood islands. If any single process corresponds to our notion of "the organizer" then it is this.

Although the dorsal-ventral gradient had clearly been described by Yamada (1937), his experiments were on neurulae and the results were not generally held to be relevant to the organizer. Dalcq and Pasteels' model (Dalcq & Pasteels, 1937) involved a dorsoventral gradient, but this related to the postfertilization egg and was likewise not considered really relevant. Cooke's papers had explicitly assumed a dorsoventral gradient although did not actually investigate its properties. I myself, obsessed by Yamada's old experiments, and interested to see if they worked at the gastrula stage, made dorsal-ventral combinations from early gastrulae of axolotl and *Xenopus* (Slack & Forman, 1980) (Fig. 7I). These confirmed that there was certainly some sort of dorsalizing signal active during gastrulation and emitted by the dorsal lip region. Suspicion that the dorsalizing signal was due to the *lack* of something, rather than the *presence* of something, came from various experiments showing that neuralization in *Xenopus* could be provoked by cell disaggregation (Godsave & Slack, 1989; Grunz & Tacke, 1989) (Fig. 7G). Although they had by now been forgotten, there were actually several earlier experiments which had involved partial disaggregation (Barth & Barth, 1959; Niu & Twitty, 1953; Yamada, 1950). In particular the study of Yamada had, in 1950, shown remarkable dorsalization of the ventral marginal zone of *Triturus* by brief treatment with 10 mM ammonia (pH about 12) Because of the toxicity of this treatment the effects had been ascribed to "sublethal cytolysis" although it was also an experiment involving cell disaggregation and reaggregation.

The real breakthrough came, again, from the new techniques of molecular biology. Harland, having previously found *Wnt8* from his expression screen based on the restoration of axes to UV irradiated embryos, now found another active gene product, which he called noggin (Lamb et al., 1993; Smith & Harland, 1992; Smith et al., 1993). This had exactly the properties expected for an organizer signal. The gene was expressed in the organizer region, the protein was a secreted molecule, it would dorsalize mesoderm and induce neural tube from ectoderm using the appropriate assays. Later biochemical study showed that it was a direct inhibitor of BMP, already known to be expressed in the ventrolateral region of the embryo (Zimmerman et al., 1996). This fitted the idea of the lack of something: ectoderm was specified to become epidermis because of the autocrine activity of BMPs. The organizer had neutralizing activity because it inhibited BMP function in a gradient-like manner.

At the same time, the lab of Melton conducting experiments with activin-blocking products obtained similar results (Hemmati Brivanlou & Melton, 1994; Hemmati Brivanlou et al., 1994). Among the inhibitors used was follistatin, considered at the time to be specific for activin, but later found also to be an inhibitor of BMP (Fainsod et al., 1997). At the same time the lab of de Robertis was cloning products from the organizer region by starting with mRNA from lithium-treated embryos, and subtracting maternal RNA. They found many transcription factors which turned out to have key roles in defining the properties of the organizer. They also found a secreted molecule they called chordin, which also turned out to be a BMP inhibitor (Sasai et al., 1994). Noggin, follistatin and chordin all have neuralizing activity, they are all expressed in the dorsal lip of the early gastrula, and they all antagonize BMP, which itself promotes ventral mesoderm and epidermal development. Furthermore, ablation of all three mRNAs for noggin, chordin and follistatin completely prevents the formation of the neural plate and the axial mesoderm (Khokha et al., 2005). This makes a good case that neural induction/dorsalization is a single process and is normally caused by a cocktail of BMP inhibitors. The neuralizing effect of cell disaggregation could then be understood as a loss of BMP activity from the tissue, because of actual loss of the factor to the medium, the loss of extracellular material necessary for its presentation to receptors, or the disruption of receptor signaling. The idea arose that neural development was the "default pathway" for the ectoderm, and that formation of epidermis itself required an inductive signal, i.e. BMP (Hemmati-Brivanlou & Melton, 1997).

Of these BMP inhibitor genes, *chordin* was of considerable evolutionary interest as it is a homolog of *Drosophila short gastrulation (sog)*, and the chordin-BMP system in vertebrates is actually homologous to the *Decapentaplegic-sog* system in insects. Although insects have the BMP homolog active dorsally and vertebrates have it active ventrally, this does indicate that dorsoventral patterning system is of extreme antiquity and is actually part of the ancestral constellation of early gene expression of the primordial triploblastic animal (DeRobertis & Sasai, 1996; Nublerjung & Arendt, 1994).

9.4 Antero-posterior patterning

The remarkable capacity of the organizer sequentially to generate inducers of head, trunk and tail had gripped the imagination since it was first discovered (Mangold, 1929, 1933; Spemann, 1931). The two gradient models of neural induction, especially that of Toivonen and Saxen (Fig. 8B), provided an explanation, at the cost of muddling the process of mesoderm induction in the blastula with posteriorization in the neural plate. It turns out that there is in fact a posterior to anterior gradient, but it is of Wnt and FGF, not of the nodal-related factors which are the endogenous mesoderm inducing factors. These gradients are responsible for patterning the trunk-tail region, and some entirely different factors are involved in the induction of the head (Fig. 12B).

Evidence for the role of FGFs in posterior-anterior patterning comes from several sources. The expression domains of several FGFs commence around the blastopore and are in a posterior location during gastrulation and neurulation (Isaacs et al., 1992). The same is true of the domain of ERK signaling, visualized with an antibody to phosphorylated ERK, which lies on the signal transduction pathway activated by the FGFs (Christen & Slack, 1999). The FGFs activate expression of genes of the CDX family which in turn activate expression of posterior HOX genes (Pownall et al., 1996). Inhibition of FGF signaling with a dominant negative receptor suppresses the formation of posterior parts but has little effect on the head (Amaya et al., 1991). Overexpression of FGFs or of CDX family genes, suppresses anterior parts and expands the size of posterior parts. A very similar story can be told for Wnt signaling (Kiecker & Niehrs, 2001; McGrew et al., 1997). The conclusion is that there is a posterior to anterior gradient of FGFs and Wnts which together activate expression of CDX and thereby HOX genes. These gradients correspond to the "transforming" activity of Nieuwkoop, or the "M factor" of Toivonen, the only difference being that the FGF/Wnt gradient seems to pattern the posterior mesoderm and endoderm as well as the neural plate.

The head is induced by a separate mechanism, and the relevant inducing factors were all discovered by the judicious application of molecular biology. Many genes active in the organizer region were identified by differential screening of cDNA libraries. Among these were *Cerberus*

(Bouwmeester et al., 1996) and *Dickkopf* (Glinka et al., 1998). *Cerberus* is expressed in the deep tissue under the dorsal lip, which becomes the anterior part of the gut. It has neural inducing activity and, when the mRNA is injected into ventral blastomeres, it will induce heads without trunk or tail. Cerberus protein was found to be an antagonist of all three factors: BMP, Wnt and nodal (Belo et al., 2000).

Dickkopf was discovered using an ingenious expression screen which tested for inclusion of a head in secondary axes formed after ventral injection of mRNA for dominant negative BMP receptor. This normally has a neuralizing effect with the induced structures not exceeding hindbrain level in completeness. *Dickkopf* is also expressed in the deep part of the organizer, centered on the prospective prechordal plate. It does not induce neural tissue from animal caps but can rescue formation of anterior axial structures in the radially symmetrical UV embryos. An endogenous head-inducing effect is supported by the fact that neutralizing antibody to Dickkopf protein suppresses anterior development. Dickkopf was found to be an inhibitor of Wnt signaling.

Both Cerberus and Dickkopf are involved in head formation and are probably supplemented by the effects of Frzb, another Wnt antagonist expressed in the organizer, which also enables head formation following overexpression of dominant negative BMP receptor (Leyns et al., 1997).

What is the origin of the anteroposterior pattern? This seems to arise from the fact that the organizer region is subdivided into two parts from the time of its formation (Zoltewicz & Gerhart, 1997). The more vegetal region becomes the deep part of the dorsal lip and expresses the head-inducing factors, and the more animal part is superficial and, along with the rest of the marginal zone, expresses Wnts and FGFs. This animal-vegetal subdivision initially arises at the time of mesoderm induction in response to the nodal-related gradient. The two regions differ in their morphogenetic behavior with the deep part invaginating by cell migration over the fibronectin layer lining the blastocoel, while the superficial part elongates by convergent-extension movements caused by cellular intercalation.

As well as genes for secreted factors, various genes for transcription factors were found to be expressed specifically in the organizer region: e.g. goosecoid (Cho et al., 1991), Lim1 (Taira et al., 1992), Not1 and 2 (von Dassow et al., 1993) (Gont et al., 1993), Siamois (Lemaire et al., 1995). The discovery of these factors defined the organizer in molecular terms. They form a gene regulatory network which both maintains the organizer and controls expression of the secreted factors.

10. How we should understand the organizer today

The organizer should really be understood as a series of developmental events which act to generate the body plan of the early embryo, starting in oogenesis and terminating by the end of neurulation. The organizer graft itself happens to involve almost all of these processes but the organizer phenomenon itself is not a single process or event.

Interestingly in his 1936 book, Spemann had written that his preferred experimental approach was.

> "..to divide the total process into larger partial processes and then to proceed to analyse each of these farther and farther....".

This seems rather at variance with his conception of the organizer as a single all dominating force. But in the end the problem was solved by methods which did break the problem down into individual processes, and, had he lived to see it, Spemann would no doubt have approved.

Since the early experiment of Waddington on avian and rabbit embryos (Waddington, 1934) there has been a desire to find a homolog of the organizer in other types of vertebrate embryos. This has been successful. Considerable study of the chick, mouse, zebrafish and even human have indicated the presence of a region which can induce a secondary axis from competent tissue (Camus & Tam, 1999; Izpisuabelmonte et al., 1993; Martyn et al., 2018; Nieto, 1999; Schier & Talbot, 1998). Many of the same gene products are expressed as in the *Xenopus* organizer and many of them have comparable biological activities in their respective organisms. There are always differences of detail between organisms that have hundreds of millions of years of evolutionary divergence between them, but a lot of the organizer processes and mechanisms do seem to be conserved across vertebrate development.

The organizer paper has in the last three decades (a remarkably long time after its initial publication) become a "citation classic". Fig. 13 shows the number of citations in each decade since publication. The slight blip in the 1930s represents the "gold rush" for the chemical basis of the organizer (or more correctly the evocator). The fact that it is just a blip reflects the minute size of the global scientific enterprise in the 1930s compared to today. The very low level of citations from the 1940s to the early 1970s indicates the obscurity into which the problem fell throughout this period. The massive peak in the 1990s represents the period of the second gold rush and the final molecular solution. The paper is still being highly cited today. This indicates a

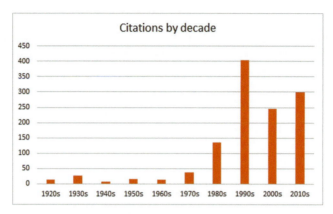

Fig. 13 Citations of the organizer paper.

persistent totemic value which this paper holds for each new generation of developmental biologists. Hopefully, those who cite it today do actually read it, do understand what it contains, and do, now, finally understand more-or-less how the organizer works.

Acknowledgments

I am grateful to Les Dale, Jim Smith and Claudio Stern for comments on the manuscript.

References

Agius, E., Oelgeschlager, M., Wessely, O., Kemp, C., & De Robertis, E. M. (2000). Endodermal Nodal-related signals and mesoderm induction in Xenopus. *Development (Cambridge, England), 127,* 1173–1183.

Amaya, E., Musci, T. J., & Kirschner, M. W. (1991). Expression of a dominant negative mutant of the FGF receptor disrupts mesoderm formation in *Xenopus* embryos. *Cell, 66* 257-250.

Asashima, M., Nakano, H., Shimada, K., Kinoshita, K., Ishi, K., Shibai, H., & Ueno, N. (1990). Mesoderm induction in early amphibian embryo by activin A (erythroid differentiation factor). *Wilhelm Roux's Archives of Developmental Biology, 198,* 330–335.

Bäckström, S. (1954). Morphogenetic effects of lithium on the embryonic development of Xenopus. *Arkiv för Zoologi, 6,* 527–536.

Balinsky, B. I. (1960). *An introduction to embryology.* Philadelphia: Saunders.

Barth, L. G., & Barth, L. J. (1959). Differentiation of cells of the Rana-pipiens gastrula in unconditioned medium. *Journal of Embryology and Experimental Morphology, 7,* 210–222.

Bautzmann, H. (1929). Über Induktion durch vordere und hintere Chorda der Neurula in verscheidenen Regionen des Wirtes. *Wilhelm Roux' Archiv für Entwicklungsmechanik der Organismen, 119,* 1–46.

Bautzmann, H., Holtfreter, J., Spemann, H., & Mangold, O. (1932). Versuche zur Analyse der Induktionsmittel in der Embryonalentwicklung. *Die Naturwissenschaften, 20,* 971–974.

Belo, J. A., Bachiller, D., Agius, E., Kemp, C., Borges, A. C., Marques, S., ... De Robertis, E. M. (2000). Cerberus-like is a secreted BMP and nodal antagonist not essential for mouse development. *Genesis (New York, N. Y.: 2000), 26,* 265–270.

Berrill, N. J., & Karp, G. (1976). *Development*. New York: McGraw Hill.

Boterenbrood, E. C., & Nieuwkoop, P. D. (1973). The formation of the mesoderm in urodelan amphibians.V. its regional induction by the endoderm. *Wilhelm Roux's Archives of Developmental Biology, 173*, 319–332.

Bouwmeester, T., Kim, S. H., Sasai, Y., Lu, B., & DeRobertis, E. M. (1996). Cerberus is a head-inducing secreted factor expressed in the anterior endoderm of Spemann's organizer. *Nature, 382*, 595–601.

Browder, L. W. (1980). *Developmental biology*. Philadelphia: Saunders.

Camus, A., & Tam, P. P. L. (1999). The organizer of the gastrulating mouse embryo. In RA Pedersen, & GP Schatten (Vol. Eds.), *Current topics in developmental biology. Vol. 45. Current topics in developmental biology* (pp. 117–153). Elsevier.

Child, C. M. (1928). The physiological gradients. *Protoplasma, 5*, 447–476.

Cho, K. W. Y., Blumberg, B., Steinbeisser, H., & de Robertis, E. M. (1991). Molecular nature of Spemann's organizer: the role of the *Xenopus* homeobox gene goosecoid. *Cell, 67*, 1111–11120.

Christen, B., & Slack, J. M. W. (1999). Spatial response to fibroblast growth factor signalling in Xenopus embryos. *Development (Cambridge, England), 126*, 119–125.

Chuang, H.-H. (1939). Induktionsleistungen von frischen und gekochten Organteilen (Niere, Leber) nach ihrer Verpflanzung in Explantate und verschiedene Wirtsregionen von Tritonkeimen. *Wilhelm Roux' Archiv für Entwicklungsmechanik der Organismen, 139*, 556–638.

Cooke, J. (1972a). Properties of primary organization field in embryo of Xenopus-laevis 2. positional information for axial organization in embryos with 2 head organizers. *Journal of Embryology and Experimental Morphology, 28*, 13.

Cooke, J. (1972b). Properties of primary organization field in embryo of Xenopus-laevis 1. Autonomy of cell behavior at site of initial organizer formation. *Journal of Embryology and Experimental Morphology, 28*, 27.

Cooke, J. (1972c). Properties of primary organization field in embryo of Xenopus-laevis 3. Retention of polarity in cell groups excised from region of early organizer. *Journal of Embryology and Experimental Morphology, 28*, 47.

Cooke, J. (1973a). Properties of primary organization field in embryo of Xenopus-laevis 4. Pattern formation and regulation following early inhibition of mitosis. *Journal of Embryology and Experimental Morphology, 30*, 49–62.

Cooke, J. (1973b). Properties of primary organization field in embryo of Xenopus-laevis 5. Regulation after removal of head organizer, in normal early gastrulae and in those already possessing a second implanted organizer. *Journal of Embryology and Experimental Morphology, 30*, 283–300.

Curtis, A. S. G. (1960). Cortical grafting in Xenopus laevis. *Journal of Embryology and Experimental Morphology, 8*, 163–173.

Curtis, A. S. G. (1962). Morphogenetic Interactions before Gastrulation in the Amphibian, Xenopus laevis—the Cortical Field. *Journal of Embryology and Experimental Morphology, 10*, 410–422.

Dalcq, A., & Pasteels, J. (1937). Une conception nouvelle des bases physiologique de la morphogénèse. *Archives de Biologie, 48*, 669–710.

Dale, L., & Slack, J. M. W. (1987a). Fate map for the 32 cell stage of *Xenopus laevis*. *Development (Cambridge, England), 99*, 527–551.

Dale, L., & Slack, J. M. W. (1987b). Regional specification within the mesoderm of early embryos of *Xenopus laevis*. *Development (Cambridge, England), 100*, 279–295.

Dale, L., Smith, J. C., & Slack, J. M. W. (1985). Mesoderm induction in *Xenopus laevis*: A quantitative study using a cell lineage label and tissue-specific antibodies. *Journal of Embryology and Experimental Morphology, 89*, 289–313.

DeRobertis, E. M., & Sasai, Y. (1996). A common plan for dorsoventral patterning in Bilateria. *Nature, 380*, 37–40.
Ede, D. A. (1978). *An introduction to developmental biology.* Glasgow: Blackie.
Fainsod, A., Deißler, K., Yelin, R., Marom, K., Epstein, M., Pillemer, G., ... Blum, M. (1997). The dorsalizing and neural inducing gene follistatin is an antagonist of BMP-4. *Mechanisms of Development, 63*, 39–50.
Fassler, P. E., & Sander, K. (1996). Hilde Mangold (1898-1924) and Spemann's organizer: achievement and tragedy. *Roux' Archives of Developmental Biology, 205*, 323–332.
Fischer, F. G., & Wehmeier, E. (1933). Zur Kenntnis der Induktionsmittel in der Embryonalentwicklung. *Die Naturwissenschaften, 21*, 518.
Gerhart, J., Ubbles, G., Black, S., Hara, K., & Kirschner, M. (1981). A reinvestigation of the role of the grey crescent axis formation in *Xenopus laevis*. *Nature, 292*, 511–516.
Gimlich, R. L., & Cooke, J. (1983). Cell lineage and the induction of 2nd nervous systems in amphibian development. *Nature, 306*, 471–473.
Glinka, A., Wu, W., Delius, H., Monaghan, A. P., Blumenstock, C., & Niehrs, C. (1998). Dickkopf-1 is a member of a new family of secreted proteins and functions in head induction. *Nature, 391*, 357–362.
Godsave, S. F., & Slack, J. M. W. (1989). Clonal analysis of mesoderm induction. *Developmental Biology, 134*, 486–490.
Gont, L. K., Steinbeisser, H., Blumberg, B., & de Robertis, E. M. (1993). Tail formation as a continuation of gastrulation: the multiple cell populations of the *Xenopus* tailbud derive from the late balstopore lip. *Development (Cambridge, England), 119*, 991–1004.
Grant, P. (1978). *Biology of developing systems.* New York: Holt, Rinehart and Winston.
Grunz, H., & Tacke, L. (1986). The inducing capacity of the presumptive endoderm of *Xenopus laevis* studied by transfilter experiments. *Wilhelm Roux's Archives of Developmental Biology, 195*, 467–473.
Grunz, H., & Tacke, L. (1989). Neural differentiation of Xenopus-laevis ectoderm takes place after disaggregation and delayed reaggregation without inducer. *Cell Differentiation and Development, 28*, 211–217.
Ham, R. G., & Veomett, M. J. (1980). *Mechanisms of development.* St.Louis: C.V.Mosby.
Hamburger, V. (1988). *The heritage of experimental embryology.* New York: Hans Spemann and the Organizer. OUP.
Heasman, J., Crawford, A., Goldstone, K., Garner-Hamvick, P., Gumbiner, B., McCrea, P., ... Wylie, C. (1994). Overexpression of cadherins and underexpression of beta-catenin inhibit dorsal mesoderm induction in early Xenopus embryos. *Cell, 79*, 791–803.
Hemmati-Brivanlou, A., & Melton, D. A. (1997). Vertebrate neural induction. *Annual Review of Neuroscience, 20*, 43–60.
Hemmati Brivanlou, A., Kelly, O., & Melton, D. A. (1994). Follistatin, an antagonist of activin, is expressed in the Spemann organizer and displays direct neuralizing activity. *Cell, 77*, 283–295.
Hemmati Brivanlou, A., & Melton, D. A. (1992). A truncated activin receptor inhibits mesoderm induction and formation of axial structures in Xenopus embryos. *Nature, 359*, 609–614.
Hemmati Brivanlou, A., & Melton, D. A. (1994). Inhibition of activin receptor signalling promotes neuralization in Xenopus. *Cell, 77*, 273–281.
Holtfreter, J. (1933). Organisierungsstufen nach regionaler Kombination von Entomesoderm mit Ektoderm. *Biologische Zentralblatt, 53*, 404–431.
Holtfreter, J. (1934a). Der einfluss thermischer, mechanischer und chemischer eingriffe auf die induzierfähigkeit von triton-keimteilen. *Wilhelm Roux' Archiv für Entwicklungsmechanik der Organismen, 132*, 225–306.

Holtfreter, J. (1934b). Über die verbreitung induzierender substanzen und Ihre Leistungen im Triton-Keim. *Wilhelm Roux' Archiv für Entwicklungsmechanik der Organismen, 132,* 307–383.

Huxley, J. S., & De Beer, G. R. (1934). *The elements of experimental embryology.* Cambridge: Cambridge University Press.

Isaacs, H. V., Tannahill, D., & Slack, J. M. W. (1992). Expression of a novel FGF in the Xenopus embryo. A new candidate inducing factor for mesoderm formation and anteroposterior specification. *Development (Cambridge, England), 114,* 711–720.

Izpisuabelmonte, J. C., Derobertis, E. M., Storey, K. G., & Stern, C. D. (1993). The homeobox gene goosecoid and the origin of organizer cells in the early chick blastoderm. *Cell, 74,* 645–659.

Jacobson, M., & Hirose, G. (1978). Origin of the retina from both sides of the embryonic brain: A contribution to the problem of crossing over at the optic chiasma. *Science (New York, N. Y.), 202,* 637–639.

Kageura, H. (1997). Activation of dorsal development by contact between the cortical dorsal determinant and the equatorial core cytoplasm in eggs of Xenopus laevis. *Development (Cambridge, England), 124,* 1543–1551.

Kao, K. R., Masui, Y., & Elinson, R. P. (1986). Lithium induced respecification of pattern in Xenopus laevis embros. *Natre, 322,* 371–373.

Khokha, M. K., Yeh, J., Grammer, T. C., & Harland, R. M. (2005). Depletion of three BMP antagonists from Spemann's organizer leads to a catastrophic loss of dorsal structures. *Developmental Cell, 8,* 401–411.

Kiecker, C., & Niehrs, C. (2001). A morphogen gradient of Wnt/beta-catenin signalling regulates anteroposterior neural patterning in Xenopus. *Development (Cambridge, England), 128,* 4189–4201.

Klein, P. S., & Melton, D. A. (1996). A molecular mechanism for the effect of lithium on development. *Proceedings of the National Academy of Sciences of the United States of America, 93,* 8455–8459.

Krieg, P. A., & Melton, D. A. (1984). Functional messenger RNAs are produced by Sp6 in vitro transcription of cloned cDNAs. *Nucleic Acids Research, 12,* 7057–7070.

Lamb, T. M., Knecht, A. K., Smith, W. C., Stachel, S. E., Economides, A. N., Stahl, N., ... Harland, R. M. (1993). Neural induction by the secreted polypeptide noggin. *Science (New York, N. Y.), 262,* 713–718.

Larabell, C. A., Torres, M., Rowning, B. A., Yost, C., Miller, J. R., Wu, M., ... Moon, R. T. (1997). Establishment of the dorso-ventral axis in Xenopus embryos is presaged by early asymmetries in beta-catenin that are modulated by the Wnt signaling pathway. *Journal of Cell Biology, 136,* 1123–1136.

Lehmann, F. E. (1937). Mesodermisierung des praesumptiven Chordamaterials durch Einwirkung von Lithiumchlorid auf die Gastrula von Triturus alpestris. *Wilhelm Roux' Archiv für Entwicklungsmechanik der Organismen, 136,* 112–146.

Lemaire, P., Garrett, N., & Gurdon, J. B. (1995). Expression cloning of siamois, a Xenopus homeobox gene expressed in dorsal vegetal cells of blastulas and able to induce a complete secondary axis. *Cell, 81,* 85–94.

Lewis, W. H. (1907). Transplantation of the lips of the blastopore in rana palustris. *American Journal of Anatomy, 7,* 137–143.

Leyns, L., Bouwmeester, T., Kim, S. H., Piccolo, S., & DeRobertis, E. M. (1997). Frzb-1 is a secreted antagonist of Wnt signaling expressed in the Spemann organizer. *Cell, 88,* 747–756.

Mangold, O. (1923). Transplantationsversuche zur Frage der Spezifität und der Bildung der Keimblätter. *Archiv für mikroskopische Anatomie und Entwicklungsmechanik, 100,* 198–301.

Mangold, O. (1929). Experimente zur Analyse der Determination und Induktion der Medullarplatte. *Wilhelm Rouxs Arch. f. EntwMech. Orgs. 117,* 586–696.

Mangold, O. (1933). Über die Induktionsfähigkeit der verscheidenen Bezirke der neurula von Urodelen. *Die Naturwissenschaften, 21*, 761–766.

Mangold, O., & Spemann, H. (1927). Uber Induktion von Medullarplatte durch Medullarplatte im Angeren Keim. *Wilhelm Rouxs Arch. f. EntwMech. Orgs, 111*, 341–422.

Martyn, I., Kanno, T. Y., Ruzo, A., Siggia, E. D., & Brivanlou, A. H. (2018). Self-organization of a human organizer by combined Wnt and Nodal signalling. *Nature, 558*, 132.

Masui, Y. (1961). Mesodermal and endodermal differentiation of the presumptive ectoderm of Triturus gastrulae through influence of lithium ion. *Experientia, 17*, 458–459.

Mayer, B. (1935). Über das Regulations- und Induktionsvermögen der halbseitigen oberen Urmundlippe von Triton. *Wilhelm Roux' Archiv für Entwicklungsmechanik der Organismen, 133*, 518–581.

McGrew, L. L., Hoppler, S., & Moon, R. T. (1997). Wnt and FGF pathways cooperatively pattern anteroposterior neural ectoderm in Xenopus. *Mechanisms of Development, 69*, 105–114.

Nakamura, O., Takasaki, H., & Ishihara, M. (1971). Formation of the organizer by combinations of presumptive ectoderm and endoderm. *Proceedings of Japanese Acadamy, 47*, 313–318.

Nakamura, O., & Toivonen, S. (1978). *Organizer. A milestone of a half-century from Spemann.* Amsterdam: Elsevier.

Nakamure, O., & Matsuzawa, T. (1967). Differentiation capacity of the marginal zone in the morula and blastula of Triturus pyrrhogaster. *Embryologia, 9*, 223–237.

Needham, J. (1942). *Biochemistry and morphogenesis.* Cambridge: Cambridge University Press.

Nieto, M. A. (1999). Reorganizing the organizer 75 years on. *Cell, 98*, 417–425.

Nieuwkoop, P. D. (1952a). Activation and organization of the central nervous system in amphibians I. Induction and activation. *Journal of Experimental Zoology, 120*, 1–31.

Nieuwkoop, P. D. (1952b). Activation and organization of the central nervous system in amphibians II. Differentiation and organization. *Journal of Experimental Zoology, 120*, 33–81.

Nieuwkoop, P. D. (1952c). Activation and organization of the central nervous system in amphibians III. Synthesis of a new working hypothesis. *Journal of Experimental Zoology, 120*, 83–108.

Nieuwkoop, P. D. (1969). The formation of the mesoderm in urodelean amphibians I. Induction by the endoderm. *Wilhelm Roux' Archiv für Entwicklungsmechanik der Organismen, 162*, 341–373.

Nieuwkoop, P. D., & Ubbels, G. A. (1972). The formation of the mesoderm in urodelean amphibians. IV. Qualitative evidence for the purely 'ectodermal' origin of the entire mesoderm and of the pharyngeal endoderm. *Roux' Archiv für Entwicklungsmechanik der Organismen, 169*, 185–199.

Niu, M. C., & Twitty, V. C. (1953). The differentiation of gastrula ectoderm in medium conditioned by axial mesoderm. *Proceedings of the National Academy of Sciences of the United States of America, 39*, 985–989.

Nublerjung, K., & Arendt, D. (1994). Is ventral in insects dorsal in vertebrates—A history of embryological arguments favoring axis inversion in chordate ancestors. *Rouxs Archives of Developmental Biology, 203*, 357–366.

Plessow, S., Davids, M., Born, J., Hoppe, P., Loppnow-Blinde, B., Tiedemann, H., & Tiedemann, H. (1990). Isolation of a vegetalizing inducing factor after extraction with acid ethanol. Concentration-dependent inducing capacity of the factor. *Cell Differentiation and Development, 32*, 27–38.

Pownall, M. E., Tucker, A. S., Slack, J. M., & Isaacs, H. V. (1996). eFGF, Xcad3 and Hox genes form a molecular pathway that establishes the anteroposterior axis in Xenopus. *Development (Cambridge, England), 122*, 3881–3892.

Roël, G., Hamilton, F. S., Gent, Y., Bain, A. A., Destrée, O., & Hoppler, S. (2002). Lef-1 and Tcf-3 transcription factors mediate tissue-specific Wnt signaling during xenopus development. *Current Biology, 12*, 1941–1945.

Sasai, Y., Lu, B., Steinbeisser, H., Geissert, D., Gont, L. K., & Derobertis, E. M. (1994). Xenopus chordin—A novel dorsalizing factor-activated by organizer-specific homeobox genes. *Cell, 79*, 779–790.

Saxén, L. (1961). Transfilter neural induction of amphibian ectoderm. *Developmental Biology, 3*, 140–152.

Saxen, L., & Toivonen, S. (1962). *Primary embryonic induction*. London: Logos Press.

Scharf, S. R., & Gerhart, J. C. (1983). Axis determination in eggs of Xenopus laevis: A critical period before first cleavage, identified by the commen effects of cold, pressure and ultraviolet irradiation. *Developmental Biology, 99*, 75–87.

Schier, A. F., & Talbot, W. S. (1998). The zebrafish organizer. *Current Opinion in Genetics & Development, 8*, 464–471.

Slack, J. M. W. (1987). Morphogenetic gradients—Past and present. *Trends in Biochemical Sciences, 12*, 200–204.

Slack, J. M. W., Darlington, B. G., Heath, J. K., & Godsave, S. F. (1987). Mesoderm induction in early Xenopus embryos by heparin-binding growth factors. *Nature, 326*, 197–200.

Slack, J. M. W., & Forman, D. (1980). An interaction between dorsal and ventral regions of the marginal zone in early amphibian embryos. *Journal of Embryology and Experimental Morphology, 56*, 283–299.

Smith, J. C. (1987). A mesoderm inducing factor is produced by a Xenopus cell line. *Development (Cambridge, England), 99*, 3–14.

Smith, J. C., Price, B. M. J., Vannimmen, K., & Huylebroeck, D. (1990). Identification of a potent Xenopus mesoderm-inducing factor as a homolog of activin-a. *Nature, 345*, 729–731.

Smith, J. C., & Slack, J. M. W. (1983). Dorsalization and neural induction: Properties of the organizer in Xenopus laevis. *Journal of Embryology and Experimental Morphology, 78*, 299–317.

Smith, W. C., & Harland, R. M. (1991). Injected Xwnt-8 Rna acts early in xenopus embryos to promote formation of a vegetal dorsalizing center. *Cell, 67*, 753–765.

Smith, W. C., & Harland, R. M. (1992). Expression cloning of noggin, a new dorsalizing factor localised to the Spemann organizer in Xenopus embryos. *Cell, 70*, 829–840.

Smith, W. C., Knecht, A. K., Wu, M., & Harland, R. M. (1993). Secreted noggin mimics the Spemann organizer in dorsalizing Xenopus mesoderm. *Nature, 361*, 547–549.

Sokol, S., Christian, J. L., Moon, R. T., & Melton, D. A. (1991). Injected Wnt RNA induces a complete body axis in Xenopus embryos. *Cell, 67*, 741–752.

Spemann, H. (1918). Über die Determination der ersten Organanlagen des Amphibienembryo I–VI. *Archiv für Entwicklungsmechanik der Organismen, 43*, 448–555.

Spemann, H. (1921). Die Erzeugung tierischer Chimären durch heteroplastische embryonale Transplantation zwischen Triton cristatus und taeniatus. *Archiv für Entwicklungsmechanik der Organismen, 48*, 533–570.

Spemann, H. (1924). Über Organisatoren in der tierischen Entwicklung. *Die Naturwissenschaften, 12*, 1092–1094.

Spemann, H. (1931). Über den Anteil von Implantat und Wirtskeim an der Orientierung und Beschaffenheit der induzierten Embryonalanlage. *Wilhelm Rouxs Archiv für Entwicklungsmechanik der Organismen, 123*, 389–517.

Spemann, H. (1938). *Embryonic development and induction.* Garland. Reprinted 1967 Hafner NY; reprinted 1988.
Spemann, H., & Mangold, H. (1924). Über Induktion von Embryonalanlagen durch Implantation artfremder Organisatoren. *Archiv für mikroskopische Anatomie und Entwicklungsmechanik, 100,* 599–638.
Taira, M., Jamrich, M., Good, P. H., & Dawid, I. B. (1992). The lim domain containing homeobox xlim1 is expressed specifically in the organizer region of xenopus gastrula embryos. *Genes & Development, 6,* 356–366.
Tiedemann, H., Lottspeich, F., Davids, M., Knöchel, S., Hoppe, P., & Tiedemann, H. (1992). The vegetalizing factor A member of the evolutionarily highly conserved activin family. *FEBS Letters, 300,* 123–126.
Toivonen, S. (1953). Bone-marrow of the Guinea-pig as a Mesodermal Inductor in Implantation Experiments with Embryos of Triturus. *Development (Cambridge, England), 1,* 97–104.
Toivonen, S., & Saxen, L. (1955). Ober die induktion des Neuralrohrs bei Tritonkeimen als simultane Leistung des Leber- und Knochenmarkgewebes von Meerschweinchen. *Annales Academiae Scientiarum Fennicae Mathematica, 30,* 1–29.
Von Dassow, C., Schmidt, J. E., & Kimelman, D. (1993). Induction of the Xenopus organizer: Expression and regulation of Xnot, a novel FGF and activin reulated homeobox gene. *Genes & Development, 7,* 355–366.
Waddington, C. H. (1933). Induction by the endoderm in birds. *Wilhelm Roux' Archiv für Entwicklungsmechanik der Organismen, 128,* 502–521.
Waddington, C. H. (1934). Experiments on embryonic induction Part I. The competence of the extra-embryonic ectoderm in the chick Part II. Experiments on coagulated organisers in the chick Part III. A note on inductions by chick primitive streak transplanted to the rabbit embryo. *Journal of Experimental Biology, 11,* 211–227.
Waddington, C. H. (1956). *Principles of embryology.* London: George Allen & Unwin.
Waddington, C. H., & Gray, J. (1932). III. Experiments on the development of chick and duck embryos, cultivated in vitro. *Philosophical Transactions of the Royal Society of London. Series B. Containing Papers of a Biological Character, 221,* 179–230.
Waddington, C. H., Needham, J., Nowinski, W. W., Needham, D. M., & Lemberg, R. (1934). Active principle of the amphibian organisation centre. *Nature, 134,* 103.
Weiss, P. (1925). Unabhängigkeit der Extremitätenregeneration vom Skelett (bei Triton cristatus). *Archiv für mikroskopische Anatomie und Entwicklungsmechanik, 104,* 359–394.
Wielstra, B. (2019). Triturus newts. *Current Biology, 29,* R105–R119.
Wolpert, L. (1969). Positional information and the spatial pattern of cellular differentiation. *Journal of Theoretical Biology, 25,* 1–47.
Yamada, T. (1937). Das Determinationzustand des Rumpfmesoderms im Molchkeim nach der Gastrulation. *Wilhelm Roux's Archiv für Entwicklungsmechanik der Organismen, 137,* 151–270.
Yamada, T. (1940). Beeinflussung der Differenzierungsleistung des isolierten mesoderms von Molchkeim durch zugefugtes Chorda- und Neural-material. *Okajima's Folia Anatomica Japonica, 19,* 131–197.
Yamada, T. (1950). Dorsalization of the ventral marginal zone of the Triturus gastrula.1. Ammonia treatment of the medio ventral marginal zone. *Biological Bulletin, 98,* 98–121.
Yamaguchi, Y., & Shinagawa, A. (1989). Marked alteration at midblastula transition in the effect of lithium on formation of larval body pattern of Xenopus laevis. *Development, Growth & Differentiation, 31,* 531–541.

Yan, L., Chen, J., Zhu, X., Sun, J., Wu, X., Shen, W., ... Meng, A. (2018). Maternal Huluwa dictates the embryonic body axis through β-catenin in vertebrates. *Science (New York, N. Y.), 362*, eaat1045.

Zimmerman, L. B., DeJesusEscobar, J. M., & Harland, R. M. (1996). The Spemann organizer signal noggin binds and inactivates bone morphogenetic protein 4. *Cell, 86*, 599–606.

Zoltewicz, J. S., & Gerhart, J. C. (1997). The Spemann organizer of Xenopus is patterned along its anteroposterior axis at the earliest gastrula stage. *Developmental Biology, 192*, 482–491.

CHAPTER TWO

The organizer and neural induction in birds and mammals

Claudio D. Stern[*]

Department of Cell and Developmental Biology, University College London, London, United Kingdom
[*]Corresponding author. e-mail address: c.stern@ucl.ac.uk

Contents

1. Historical introduction	44
2. Hensen's node ("the node")	44
2.1 Cellular composition of the node	45
3. Neural induction: spatial and temporal aspects	47
4. Molecular mechanisms of neural induction	49
5. The organizer and "dorso-ventral" (axial-lateral) patterning	53
6. Neural induction and rostro-caudal (anterior-posterior) patterning: how many organizers?	53
7. Is neural induction by a grafted organizer comparable to normal neural plate development?	56
8. Neural induction in vivo and in vitro	58
9. Are the mechanisms of neural induction in amniotes different from those in anamniotes?	58
References	60

Abstract

In avian and mammalian embryos the "organizer" property associated with neural induction of competent ectoderm into a neural plate and its subsequent patterning into rostro-caudal domains resides at the tip of the primitive streak before neurulation begins, and before a morphological Hensen's node is discernible. The same region and its later derivatives (like the notochord) also have the ability to "dorsalize" the adjacent mesoderm, for example by converting lateral plate mesoderm into paraxial (pre-somitic) mesoderm. Both neural induction and dorsalization of the mesoderm involve inhibition of BMP, and the former also requires other signals. This review surveys the key experiments done to elucidate the functions of the organizer and the mechanisms of neural induction in amniotes. We conclude that the mechanisms of neural induction in amniotes and anamniotes are likely to be largely the same; apparent differences are likely to be due to differences in experimental approaches dictated by embryo topology and other practical constraints. We also discuss the relationships between "neural induction" assessed by grafts of the organizer and normal neural plate development, as well as how neural induction relates to the generation of neuronal cells from embryonic and other stem cells in vitro.

1. Historical introduction

This year sees the 100th anniversary of the publication of the influential paper by Spemann and Mangold, reporting that a small region of the vertebrate embryo, named the "organizer", can induce ectodermal cells that do not normally contribute to the neural plate to form a complete, patterned nervous system (Spemann & Mangold, 1924; Spemann, 1921). In amphibians, where these experiments were initially conducted, the "organizer" resides in the dorsal lip of the blastopore (see accompanying review by Slack, 2024). A few years later, Waddington demonstrated that an equivalent region exists in birds (duck and chick) and mammals (rabbit) (Waddington & Schmidt, 1933; Waddington, 1933, 1934, 1936, 1937): the tip of the primitive streak. Later work extended this to the mouse embryo (Beddington, 1994; Camus & Tam, 1999; Kinder et al., 2001; Robb & Tam, 2004).

This interaction between the organizer and the responding ectoderm, which causes the latter to acquire neural plate identity, has been termed "neural induction" (Gallera, 1971b; Gurdon, 1987; Nieuwkoop et al., 1952; Saxen, 1980; Spemann & Mangold, 1924; Stern, 2005; Storey, Crossley, De Robertis, Norris, & Stern, 1992; Trevers et al., 2023), but signals from the organizer also influence other tissues, especially axial-lateral (called "dorsoventral" in the amphibian and fish literature) cell identities (somites, intermediate mesoderm, lateral plate) (Slack, 2024). This review surveys historical and current knowledge about the organizer and its functions, including neural induction, in amniotes. Most is known about this in chick and mouse, although new information is starting to emerge about the equivalent region in human embryos.

2. Hensen's node ("the node")

At the full primitive streak stage (Hamburger-Hamilton [HH] stage 4, just prior to the first emergence of precursors for the head process, prechordal mesendoderm and notochord at HH4+) (Hamburger & Hamilton, 1951), the tip of the primitive streak becomes a bulbous thickening. This was first explicitly described by Viktor Hensen in embryos of the guinea pig and rabbit (Hensen, 1876), who described it as a knot (*Knoten* in German). By extension (even though a bulbous node is not yet present at earlier stages), the tip of the advancing primitive streak of avian and many mammalian embryos is referred to as "Hensen's node". Probably with the mistaken assumption that the original description of this structure by Viktor Hensen was made in chick (rather than two mammalian species), a consensus reached at a discussion meeting in London in 1991 decided to name the mouse equivalent structure as "the node" (Multi-author, 1992). As we shall see later, however, by the time a visible bulge or node appears, the tip of the primitive streak has largely lost its neural inducing ability.

2.1 Cellular composition of the node

The primitive streak arises during gastrulation as a midline thickening of the epiblast, which lengthens over a period of hours in a posterior-to-anterior (caudal to cranial) direction (Fig. 1). Through it, epiblast cells ingress to give rise to mesoderm and endoderm, over a prolonged period of time. After its elongation, and following from the appearance of the node as a visible structure, the primitive streak starts to shorten ("regression", from HH6) and eventually becomes incorporated into the tail bud (Bellairs, 1986). Several studies in the chick embryo have revealed that the tip of the mature primitive streak (Hensen's node) is formed from contributions from two distinct populations of cells. One of these is the midline of a crescent-shaped

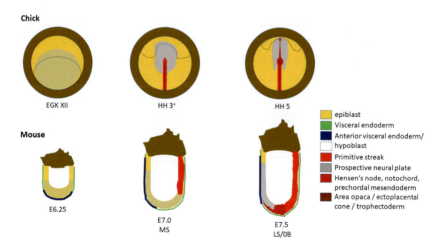

Fig. 1 Schematic diagrams of chick (*upper row*) and mouse (*lower row*) embryos at primitive streak stages. Chick embryos are shown at pre-primitive streak stage (EGK XII) (Eyal-Giladi & Kochav, 1976), mid-primitive streak (HH3[+]) and head process (early neurulation; HH5). Mouse embryos are shown at approximately E6.5 (before primitive streak formation), mid-streak (MS; approximately E7.0) and late streak/zero-bud (LS/0B; approximately E7.5) (Downs & Davies, 1993). The color key on the right indicates the tissues and regions relevant to this review.

thickening, called "Koller's sickle", at the posterior edge of the epiblast prior to primitive streak formation (Bachvarova, Skromne, & Stern, 1998; Callebaut & Van Nueten, 1994; Izpisúa-Belmonte, De Robertis, Storey, & Stern, 1993; Koller, 1882). Before streak formation, the embryonic epiblast adjacent to the sickle moves anteriorly, as part of convergence-extension movements in the posterior epiblast driven by cell intercalation (Voiculescu, Bertocchini, Wolpert, Keller, & Stern, 2007; Voiculescu, Bodenstein, Lau, & Stern, 2014), stopping close to the geometrical center of the disc-shaped embryonic epiblast. There it is reached, at HH3–3+, by the sickle-derived population whose movement accompanies elongation of the primitive streak (Izpisúa-Belmonte et al., 1993; Streit, Berliner, Papanayotou, Sirulnik, & Stern, 2000). After the two populations come together, the tip of the primitive streak continues to recruit epiblast cells that incorporate into it before some of them emerge as mesoderm or endoderm (Joubin & Stern, 1999; Psychoyos & Stern, 1996; Solovieva, Lu, Moverley, Plachta, & Stern, 2022a). Although there is as yet no information about whether the two principal precursor populations can be distinguished molecularly, neither population has full neural inducing ability until the two come together at the tip of the primitive streak (Streit et al., 2000).

Lineage tracing of single cells within the node revealed that although the majority of the cell population in the node is transient (cells enter and leave, as do their descendants, as in the rest of the primitive streak), a small subset of cells seems to reside in the node for a prolonged period while their descendants leave at each cell division to contribute to notochord and presomitic mesoderm (McGrew et al., 2008; Selleck & Stern, 1991; Selleck & Stern, 1992; Solovieva et al., 2022a). Serial transplantation of single cells suggests that the node contains a "niche" that can instruct cells to acquire this self-renewing, stem cell like behavior that can be interpreted even by cells from a remote location that would not otherwise have entered the node (Solovieva et al., 2022a). The resident stem cell niche is located at the posterior part of the node in both chick and mouse embryos, a region named "node-streak border" and persists well into neurulation stages (Cambray & Wilson, 2002; Solovieva et al., 2022a; Solovieva, Wilson, & Stern, 2022b), unlike the recruitment of cells from the lateral epiblast, which ends just before the notochord/head process emerges, around the time a morphological node appears (Sheng, Dos Reis, & Stern, 2003; Solovieva et al., 2022a). Single cell RNA-sequencing of the resident cells and their neighbors revealed some heterogeneity between these populations; indeed in situ hybridization reveals that some transcripts are expressed in a mosaic (salt-and-pepper) fashion in the node. Therefore, although the entire node population expresses distinctive genes such as Goosecoid, Chordin (Izpisúa-Belmonte et al., 1993; Streit et al., 1998) and others, there are molecularly distinct subpopulations of cells which may well have different functions. The question of whether the "organizer" activity (including neural induction and patterning, and mesoderm patterning) is a property of all cells in the node or whether it requires cooperative action of different cell populations within it has not yet been addressed.

3. Neural induction: spatial and temporal aspects

Since the chick embryo is a large flat disc, transplantation and graft removal experiments can be done at a wide range of stages into any region of the embryo, to a degree that is not possible in amphibians, fish or mouse. Along with this, the introduction of the quail/chick chimaera technique in the 1960s allowed unambiguous identification of the sources of cells (donor- or host-derived) contributing to different structures. This has allowed detailed information to emerge concerning both the signaling ability of the anterior primitive streak/node, and the spatial and temporal constraints (competence) of the responding epiblast. Grafts of the node/anterior streak to the inner area pellucida of a host embryo generate an ectopic axis, often fused to that of the host, but it is not possible to determine whether the cells contributing to the ectopic neural plate are "recruited" from those that would normally have contributed to the normal host neural plate. However, a large part of the more peripheral area opaca (extraembryonic epiblast) is competent to respond to organizer grafts and generate a complete, fully patterned neural plate derived entirely from the host (Dias & Schoenwolf, 1990; Gallera & Ivanov, 1964; Gallera, 1971a; Storey et al., 1992) (Fig. 2).

Using donor embryos at different stages as the source of the grafted anterior streak/node revealed that the tip of the primitive streak between stages HH3–4 has the ability to induce a fully patterned neural plate from host area opaca epiblast (hosts at stage HH3+). The node from stage HH4+ has lower ability to induce (down to about 50% of cases), but the graft itself contributes more to the ectopic neural plate, suggesting that at these later stages the node area contains some specified prospective neural plate cells (Dias & Schoenwolf, 1990; Storey et al., 1992). The epiblast of the area opaca of the host, in turn, loses its competence completely,

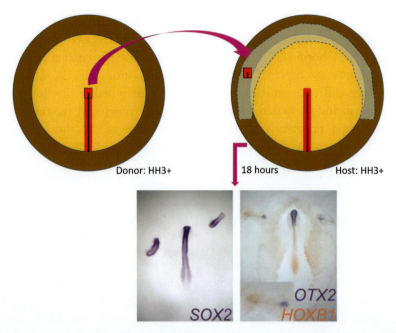

Fig. 2 Diagram showing the standard assay for neural induction in the chick embryo. The epiblast, area opaca and primitive streak are shown in the same color scheme as in Fig. 1. The competent portion of the area opaca and outer non-neural epiblast are shown outlined with a dashed line in the host embryo (*right*). The anterior tip of the primitive streak of a donor embryo (which can be a quail or a GFP-transgenic embryo) is transplanted adjacent to the epiblast within the competent region of the host; both donor and host are at mid-primitive streak stage (HH3$^+$). The lower images show two embryos that had received two grafts each (one on each side), after overnight culture (about 18 h) and in situ hybridization for Sox2 to mark the entire neural plate (*left*) or Otx2 (*blue*) and HoxB1 (*brick color*) to mark the forebrain and hindbrain/spinal cord, respectively (*right*). Lower images from (Trevers et al., 2023).

immediately after HH4 (just after a morphological Hensen's node appears) (Dias & Schoenwolf, 1990; Gallera & Ivanov, 1964; Storey et al., 1992). Within the area opaca, competence is restricted to the inner one-third to one-half of the anterior half of this region (Dias & Schoenwolf, 1990; Gallera & Ivanov, 1964; Storey et al., 1992; Streit et al., 1997).

Timed transplantation experiments revealed that the responding ectoderm requires about 12 h' exposure to the organizer (Gallera & Ivanov, 1964; Gallera, 1971b; Streit et al., 1998; Trevers et al., 2023) to acquire neural identity in a stable way, even after the organizer graft has been removed ("commitment"). Recent work has also revealed that the expression of many genes changes over time after grafting an organizer, suggesting that the process has considerable complexity (Albazerchi & Stern, 2007; Gibson, Robinson, Streit, Sheng, & Stern, 2011; Papanayotou et al., 2008; Papanayotou et al., 2013; Pinho et al., 2011; Sheng et al., 2003; Stern, 2005; Streit & Stern, 1999a; Streit et al., 1997; Streit et al., 1998; Streit et al., 2000; Trevers et al., 2018; Trevers et al., 2023). The initial responses (the first 1–5 h after exposure to the graft) are characterized by induction of a transcriptional signature that resembles the border of the neural plate and also early embryonic epiblast and embryonic stem cells (Lavial et al., 2007; Trevers et al., 2018; Trevers et al., 2023). Indeed, if epiblast that has been exposed to signals from a graft of anterior primitive streak for 5 h is explanted into culture, it develops a neural plate border-like identity, expressing pre-placodal markers (Trevers et al., 2018).

Interestingly, the initial responses to a graft of the "organizer" (anterior primitive streak) and to other inducing tissues such as the lateral head mesoderm (placode-inducing) or the hypoblast (see below) are very similar in the first 5 h after the graft. The responses begin to diverge after 5 h. Neural induction can be elicited by first exposure to any of these 3 tissues, followed by replacement by an anterior primitive streak graft, and likewise placode induction can be obtained by first exposure to any of the tissues followed by replacement by the lateral head mesoderm (Trevers et al., 2018). This led to the idea of a "common state" underlying the earliest stages of induction in response to any inducing tissue to which the epiblast is competent, with later signals determining the final outcome of the induction (Thiery, Buzzi, & Streit, 2020; Trevers et al., 2018; Trevers et al., 2023).

4. Molecular mechanisms of neural induction

Classically, mainly from the amphibian literature since Spemann and Mangold, "neural induction" has been imagined as a discrete "switch" between a non-neural state and a neural state of the responding cells. Most of the experiments involved some manipulation at early gastrula or blastula stage or even before (early cleavage), and analysis of the outcome more than a day later, at neurulation stages or later. These experiments did not establish when the "decision" between non-neural and neural takes place within this long window. Insight into this came from experiments in the chick where the organizer was removed after different times following grafting—this approach established that 12–13 h' contact between competent responding epiblast and the primitive streak/node graft are required for the former then to continue to develop into a morphological neural plate and express appropriate definitive neural plate markers including Sox1 (Gallera, 1965, 1971b; Streit et al., 1998). A recent study expanded on this by analyzing the transcriptional responses to an organizer (anterior primitive streak) graft in fine time-course (1–2 h intervals after the graft) and identified thousands of genes, including more than 300 transcriptional regulators, whose expression changes at some point in response to signals from the organizer. Chromatin immunoprecipitation with sequencing (ChIP-seq), analysis of transposase-sensitive sites (ATAC-seq) and NanoString transcriptional analysis, together with bioinformatics approaches were then used in a computational pipeline to construct a detailed Gene Regulatory Network (GRN) of responses in time-course between first exposure to the organizer (0 h) and "commitment" to a neural plate identity at 12–13 h (Trevers et al., 2023). The network comprises 175 transcriptional regulators and 5614 predicted interactions between them, in fine time-course. It shows progressive increase in complexity of the responses and the transition between the early "common state" (see above) and the acquisition of a neural plate fate (Fig. 3).

What about the signals emitted from the organizer that input into (and thus regulate) this network? In amphibians (and subsequently in fish), the dominant view for the last 3 decades has been that a single input, namely inhibition of BMP signaling, is sufficient to assign "dorsal" fates to competent ectoderm cells, and that these dorsal fates include neural plate (reviewed in the accompanying chapter by Slack, 2024). Key experiments by many groups in the 1990s established that endogenous secreted BMP inhibitors including Chordin, Noggin and Follistatin are expressed in the Spemann organizer and act on neighboring ectodermal cells, inhibiting the "ventralizing" properties of BMPs which are expressed ubiquitously in ectoderm at the blastula stage; this led to the "default model", which proposes that ectodermal cells deprived of any signal (including BMP) are specified as neural "by default" (Harland & Gerhart, 1997; Hemmati-Brivanlou & Melton, 1997; Lamb et al., 1993; Piccolo, Sasai, Lu, & De Robertis, 1996; Sasai et al., 1994; Sasai, Lu, Steinbeisser, & De Robertis, 1995; Thomsen, 1995; Wilson & Hemmati-Brivanlou, 1995; Zimmerman, De Jesus-Escobar, & Harland, 1996). However, experiments mainly in the chick, but also some in Xenopus, challenge this simplistic view.

First, it was found that the major neural inducers identified in Xenopus are expressed in the chick organizer at the appropriate time to account for neural induction: Chordin is expressed early in the anterior primitive

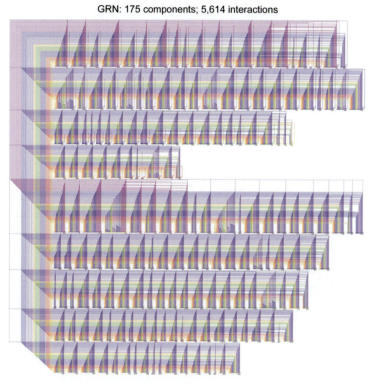

Fig. 3 A gene regulatory network for neural induction, comprising 175 transcriptional regulators and 5614 predicted interactions between them. *From Trevers, K. E., Lu, H. C., Yang, Y., Thiery, A. P., Strobl, A. C., Anderson, C., ... Stern, C. D. (2023). A gene regulatory network for neural induction. Elife, 12, e73189.*

streak, but its expression persists, very strongly, at a stage when the organizer has virtually lost all inducing activity (even at tail bud stages) (Streit et al., 1998). Noggin is not expressed in the organizer at all, at any stage—it is expressed in the emerging head process/notochord but these do not have neural inducing activity (Streit & Stern, 1999a,b). Follistatin is also not expressed in the organizer at any stage. Moreover, none of these inhibitors, either alone or in combination, or any other means of inhibiting BMP such as overexpression of the inhibitory smad SMAD7, or of dominant-negative BMP receptors, or combinations of many of the above, are able to induce neural plate markers in competent epiblast that can respond to the organizer (de Almeida et al., 2008; Linker & Stern, 2004; Streit et al., 2000). However, when BMP antagonists are misexpressed as a line that crosses the prospective neural plate border, they are able to move this border outwards into non-neural territories of the epiblast, and thus expand the neural plate (Linker et al., 2009). Interestingly, the same is found in Xenopus (Linker et al., 2009), and detailed fate maps reveal that even "small" animal caps, which are often used in neural induction assays in Xenopus, in fact contain precursors for the neural crest, a neural plate border derivative (Linker et al., 2009), suggesting that animal cap assays are biased and do not test neural inducing activity on truly non-neural cells. A more appropriate assay would be to use ventral ectoderm (presumptive epidermis) but these experiments are not usually performed. To note, the animal cap assay was originally developed to test mesoderm induction (because the animal pole does not contain any mesodermal precursors) by Nieuwkoop (Nieuwkoop, 1969a, 1969b) and only later adopted for neural induction.

Fibroblast growth factors quickly emerged as another important player in both chick (Akai & Storey, 2003; Alvarez, Araújo, & Nieto, 1998; Rodríguez-Gallardo, Climent, Garcia-Martinez, Schoenwolf, & Alvarez, 1997; Storey et al., 1998; Streit & Stern, 1999a; Streit et al., 2000) and Xenopus—in the latter, inhibition of FGF prevents the neural inducing ability of both Chordin (Sasai, Lu, Piccolo, & De Robertis, 1996) and Noggin (Delaune, Lemaire, & Kodjabachian, 2005; Launay, Fromentoux, Shi, & Boucaut, 1996). In chick, timed organizer grafts and removal experiments followed by application of BMP antagonists suggested that although BMP inhibition is indeed required for neural induction, cells need to be exposed to organizer signals for at least 5 h (presumably to induce the "common state") before they can respond to BMP antagonists (Linker & Stern, 2004; Streit et al., 2000). Together these findings suggest that at least FGF is also required for neural induction, and upstream of BMP inhibition. However, FGF exposure either together with, or followed by BMP inhibition, is still insufficient to induce neural plate markers, suggesting that yet other players are involved; Wnt inhibition has also been suggested to play a role (Wilson et al., 2001). Despite this, even a combination of FGF, BMP-inhibition and Wnt-inhibition is unable to induce neural plate markers in competent chick epiblast, suggesting that more signals, remaining to be identified, may contribute to the process. The recently described GRN for neural induction (Trevers et al., 2023) may help as a platform to test for these signals, as well as to understand the molecular basis of the mechanisms underlying the establishment and loss of competence of the epiblast. Furthermore, our recent finding that the organizer is made up of a heterogeneous population of cells raises the possibility that these distinct populations have different signaling properties which together contribute to the full functional properties of the organizer.

5. The organizer and "dorso-ventral" (axial-lateral) patterning

The classical organizer transplantation experiments in amphibians sometimes generated ectopic somites – these were mainly host-derived, and as members of the Spemann lab and others quickly realized, were the consequences of a re-specification of the host lateral plate ("ventral") mesoderm to a paraxial (pre-somitic, more "dorsal") fate (see accompanying review by Slack, 1991). This was also found in the chick, first observed by Waddington in the 1930s but more compellingly demonstrated by Nicolet (Nicolet, 1968, 1970, 1971). One experiment, for example, involved co-transplantation of a node with an explant from the *posterior* primitive streak, normally destined to give rise only to lateral plate mesoderm, to a remote position in a host embryo: under the influence of the node, the streak explant now generates somites. Eventually it became clear, as in amphibians, that the key actor in this process is BMP inhibition by factors (mainly Noggin) expressed in the node and the notochord that emerges from it (Dias, De Almeida, Belmonte, Glazier, & Stern, 2014; Streit & Stern, 1999b; Tonegawa & Takahashi, 1998; Tonegawa, Funayama, Ueno, & Takahashi, 1997). Interestingly, Chordin, another BMP inhibitor also expressed in the notochord at the same time, does not share this activity of Noggin (Streit & Stern, 1999b), suggesting different specificities in the BMP ligands inhibited by these two secreted proteins.

By the way, the terminology "dorso-ventral" for this axis of the mesoderm or indeed of the embryo as a whole (from the organizer side to the opposite equatorial position in a blastula stage Xenopus embryo), used in the amphibian and fish embryological literature, is very confusing for amniote embryologists because the equivalent regions of the mesoderm extend bilaterally from the midline of the embryonic disc (most amniotes, other than rodents, have disc-like morphology rather than spherical). A more appropriate term that is independent of embryo geometry and emphasizes the conserved features would be "axial-lateral", where "axial" refers to the axis described by the notochord and the midline of the overlying neural plate/tube.

6. Neural induction and rostro-caudal (anterior-posterior) patterning: how many organizers?

The organizer clearly has an influence on the extent of "patterning" (orderly arrangement of structures along the head-tail axis of the induced neural tube). One clear indication of this is that a graft of the tip of the primitive streak from a primitive streak stage (HH3+−4) donor will induce a complete nervous system from the host epiblast, whereas grafts of the node from older donors induce progressively truncated nervous systems, first losing the forebrain, then the truncation increasing to more posterior regions (Storey et al., 1992). However, the ability of an older node to induce more rostral structures like forebrain can be rescued by co-transplantation of anterior axial mesendoderm (prechordal mesendoderm) together with the older node (Foley, Storey, & Stern, 1997; Rowan, Stern, & Storey, 1999). These findings are consistent with the idea (see above, "Cellular composition of the node") that the anterior primitive streak might contain different populations of cells, with different inducing properties; as some of these leave the node to contribute to the destination tissues (such as prechordal mesendoderm) the node loses the ability to induce the corresponding structures.

At first sight, this finding appears to support the idea proposed by Otto Mangold that different regions of the amphibian mesoderm can induce different sub-regions of the CNS (Mangold, 1933), and which led to the idea of distinct "head", "trunk" and "tail" organizers. However, although the chick prechordal mesendoderm can both rescue the ability of an older node to induce anterior structures, and a graft of prechordal mesendoderm into the prospective hindbrain region can "anteriorize" the adjacent host tissue to form a patch of forebrain (Foley et al., 1997), this mesendoderm cannot induce neural tissue de novo from non-neural epiblast, and therefore cannot be considered a true "organizer".

Interestingly in the mouse, the conclusion that the "organizer" represents a succession of different cell populations with different inducing and patterning properties was also reached, although like in the chick, this is only through correlation between the results of transplantation experiments, gene expression patterns and fate maps (Camus & Tam, 1999; Kinder et al., 2001). Direct testing of the inducing and patterning abilities of different cell populations needs to be done in both species, which requires novel ways to purify them from the cell mixture in the anterior streak. In mouse, the successive stages of the anterior primitive streak receive different names, namely the "EGO" (early gastrula organizer), "MGO" (mid-gastrula organizer) and "node" (equivalent to the chick late streak stage [HH4+−5] Hensen's node, which in the chick has reduced inducing activity). These terms are unfortunate because to those unfamiliar with the detailed movements and fates of cells at these stages, they imply completely separate organizers. Given that the streak/node in both mouse and chick is composed of both resident and transient cell populations, it is more useful to refer to the structure and stage rather than to identify separate "organizers".

In 1996, Rosa Beddington observed that a component of the extra-embryonic visceral endoderm (called Anterior Visceral Endoderm, AVE) moves from a distal position to an anterior position in the mouse pre-primitive streak embryo. Removal of this tissue impairs rostral CNS development, and the paper concluded that *"primitive endoderm is responsible for the initial induction of rostral identity in the embryo, and in particular for the correct definition of the future prosencephalic neurectoderm. Subsequently, this identity will be reinforced and maintained by axial mesendoderm when it displaces the visceral embryonic endoderm during the course of gastrulation."* (Thomas & Beddington, 1996). Subsequent work in both mouse and chick established that rather than a direct induction of prosencephalic identity by the AVE (or its chick equivalent, the hypoblast), this tissue is important in positioning the primitive streak by inhibiting Nodal signaling and thus preventing streak formation in an anterior position. In chick, removal of the hypoblast (which expresses the Nodal antagonist Cerberus) causes multiple primitive streaks to appear at random positions in the blastodisc (Bertocchini & Stern, 2002), whereas in mouse a similar phenotype occurs when the two Nodal antagonists expressed in the AVE (Cerberus and Lefty-1) are knocked out by mutation (Perea-Gómez et al., 2002). As we discussed above, the chick hypoblast can induce "common state" genes similar to the initial responses of epiblast to signals from the node and other inducing tissues, but it does not fully induce neural tissue by itself, whether anterior or otherwise (Foley, Skromne, & Stern, 2000; Trevers et al., 2018). In mouse, the AVE by itself can also not induce neural tissue, but combination of the AVE with the EGO (early anterior primitive streak) does (Tam & Steiner, 1999).

The idea that the AVE is responsible for inducing the most anterior parts of the CNS also developed because grafts of "the node" can induce neural plate, but this lacks the most anterior regions (Beddington, 1994; Tam & Steiner, 1999; Thomas & Beddington, 1996). However, the "node" in the mouse is equivalent to Hensen's node of the chick, which arises at the tip of the primitive streak just at the time the first axial descendants (prospective prechordal and head process mesendoderm) are leaving the region (HH4+−5). At this stage, the chick node also lacks the ability to induce anterior structures and also has reduced inducing ability (see above), whereas the tip of the streak at earlier stages can. A second problem with the transplantation experiments in mouse is that because of the topology of the embryo, it is not possible to perform a graft to assess induction that is sufficiently far from the future host neural plate to be sure that the ectopic CNS is not composed of host prospective neural plate cells.

Unfortunately, most textbooks and recent papers still repeat the notion that the AVE is a direct inducer of "anterior fates" or even anterior CNS identity, even though there is no direct evidence for this. Rather, the most parsimonious interpretation of the data available in both chick and mouse, which are largely consistent with each other (Stern & Downs, 2012), suggests a different view. The hypoblast/AVE shares with the anterior primitive streak the ability to induce a set of early response genes (the "common state") but this is not sufficient to change the fate of cells. The anterior primitive streak is able to induce a complete CNS from epiblast that is separate from the future neural plate of the host, but it loses the ability to induce anterior structures at the time the morphological node appears and the first axial precursors leave the region. The main function of the hypoblast/AVE is at an earlier stage, to prevent the formation of a primitive streak (and mesoderm) other than at the most "posterior" end of the embryo rather than to specify anterior or neural identity, but it is also able to provide the initial signals required for neural induction (resulting in the "common state"), presumably because of shared signals with the anterior primitive streak.

7. Is neural induction by a grafted organizer comparable to normal neural plate development?

Although a graft of the anterior primitive streak can induce a fully patterned nervous system including forebrain, midbrain, hindbrain and spinal cord, as well as neural crest and placodes from the host epiblast of the area opaca, this is probably not how neural development proceeds in the normal embryo. First, more than half of the normal neural plate (the prospective cranial regions: forebrain, midbrain and most of the hindbrain), never appear to be in close proximity to the anterior primitive streak (Hatada & Stern, 1994; Metzis et al., 2018; Rosenquist, 1966; Sanchez-Arrones, Ferran, Rodriguez-Gallardo, & Puelles, 2009; Sanchez-Arrones, Stern, Bovolenta, & Puelles, 2012; Schoenwolf & Sheard, 1990; Schoenwolf, Bortier, & Vakaet, 1989; Schoenwolf, 1992; Uchikawa et al., 2011). This implies that signals from tissues other than the node or streak are likely to play a role, especially at the earliest stages of neural development. In the embryo, the "common state" genes are expressed very early, before primitive streak formation, in a domain overlying the area covered by the hypoblast. The hypoblast expresses FGF8, and this factor alone is able to induce expression of the "common state" genes (Streit et al., 2000; Trevers et al., 2018). Therefore, although the node can induce an entire CNS including the earliest steps (as it also expresses FGF8), this is not what happens in normal development. Current evidence suggests that in normal development, the prospective neural plate develops as a result of several inductive interactions, which may occur sequentially as well as concomitantly.

However, comparison between the time-course of gene expression (from the GRN) in response to an organizer graft into a competent non-neural region and expression of the same genes in the normal embryo at different stages can help to elucidate to what extent the two processes are similar. In situ hybridization revealed that genes whose expression is induced by a node graft within 3 h tend to be expressed in the pre-primitive streak stage embryo (around stage EGK XIII), those induced at 5–7 h are expressed in the central area pellucida anterior to and around the node at stages 4–6, and those induced at 9–12 h are expressed in the elevating neural plate from stages 7–9 (Trevers et al., 2023). The activity of several enhancers predicted from the GRN was tested using a vector (Uchikawa, Ishida, Takemoto, Kamachi, & Kondoh, 2003; Uchikawa et al., 2011) containing a minimal promoter (TK), the putative enhancer, and the coding sequence of GFP, electroporated into a wide region of the early embryo. In all cases tested, the domain of activity resembled that of the normal expression of the associated gene, and the stage at which the reporter first became active closely matched the stage at which the gene normally starts to be expressed (Trevers et al., 2023). Single cell RNA sequencing of normal neural plate cells from different stages revealed a very close correspondence between normal neural plate development and the sequence of transcriptional changes in GRN genes taking place following a node graft (Thiery et al., 2023; Trevers et al., 2023). Together, these three approaches all strongly suggest that the sequence of events resulting from a graft of the organizer into the area opaca epiblast closely follows the dynamics of gene expression in the normal neural plate.

The results discussed above suggest that in the normal embryo, it is the hypoblast rather than the node that initiates neural plate development (Foley et al., 2000; Streit et al., 2000; Trevers et al., 2018), even though the hypoblast does not (and cannot) induce a neural plate by itself. However, a node graft can mimic both these initial steps and those that follow. Evidence from the mouse (where the AVE corresponds to the avian hypoblast) is consistent with these conclusions (Beddington, 1994; Camus & Tam, 1999; Kinder et al., 2001; Robb & Tam, 2004; Tam & Steiner, 1999; Thomas & Beddington, 1996).

8. Neural induction in vivo and in vitro

A growing literature makes use of protocols to direct cultured mouse ES cells or human Induced Pluripotent Stem cells into neurons, and calls this process "neural induction". After maintaining the stem cell population in a medium containing FGF to sustain proliferation and pluripotency, a common current protocol involves withdrawal of the FGF, followed by activation of the hedgehog pathway (Gouti et al., 2014; Metzis et al., 2018). Other protocols in the literature include BMP inhibition, manipulation of the Wnt pathway and/or retinoid signaling, in various combinations. Many of these protocols need at least 5 days, and sometimes as long as 14, for the culture to generate neuronal cells. This is a much longer period than we see in vivo (in mouse, human or chick) either in normal development or after an organizer graft, which raises the question of to what extent they represent equivalent processes. Given that an organizer graft mimics the time-course of events of normal development, but the latter cannot be entirely ascribed to functions of the organizer (anterior primitive streak/node) (see previous section), and also that the term "induction" requires that the cells responding to the interaction should have a fate other than the results of the induction (Gurdon, 1987), it seems more prudent to reserve the term "neural induction" for events that result after an organizer graft into a non-neural domain in the embryo.

9. Are the mechanisms of neural induction in amniotes different from those in anamniotes?

In amphibians, the view that BMP inhibition alone provides a complete and sufficient explanation for neural induction remains dominant, despite the existence of strong evidence for other signals being not only involved, but also required for neural induction. Part of the reason for these not being taken sufficiently into account is the notion that FGF signaling (as well as IGF) can act as a BMP inhibitor by phosphorylating a linker region in Smad1 (Pera, Ikeda, Eivers, & De Robertis, 2003). There are two additional problems with the literature on neural induction in Xenopus: first, that most experiments rely on manipulations of BMP signaling on very early (pre-blastula) stage embryos; at this stage, BMP inhibition strongly dorsalizes the entire embryo, generating a near-circular organizer-like region around the equator. Most of the vegetal part of the embryo, including the equatorial marginal zone, are then cut out and discarded, and assays only involve assessment of the state of the animal pole at a much later stage. Therefore the early effects on the extent of the organizer are hidden. Second, many assays use isolated animal caps. Even in those cases where these are very small, they are likely to include a region whose normal fate is neural plate border/neural crest, and even perhaps some of the endogenous neural plate (Linker et al., 2009). Ideally, organizer grafts and molecular manipulations should be done in a region that is entirely non-neural in fate, to assess the "change of fate" that is crucial to the concept of an inductive interaction (Gurdon, 1987). To my knowledge this has not been done. Mouse embryos have a similar problem, because their cylindrical topology means that many regions lie very close to each other and there is a negligibly small non-neural ectoderm and the extraembryonic ectoderm is compacted at the top of the cylinder, but all regions are also close to the posterior primitive streak. Among the mammals, the rabbit embryo is probably the best to use for experimentation, but apart from the pioneering experiments of Waddington in the 1930s (Waddington & Waterman, 1933; Waddington, 1937) and more recent ones from the group of Christoph Viebahn (Hopf, Viebahn, & Puschel, 2011; Knoetgen, Teichmann, Wittler, Viebahn, & Kessel, 2000), this has not been exploited. Other flat embryos like the marmoset could allow such experiments too, but few groups use this model at present. In humans, the stages at which these experiments need to be done extend beyond the 14 day post-fertilization that is the current limit for experimentation on human embryos. For these reasons the chick remains the best model to dissect the events of neural induction and the time-course of responses.

But are there differences between amniotes and anamniotes? One obvious difference is the existence of an extraembryonic "endoderm" (hypoblast/visceral endoderm) in amniotes which does not exist in anamniote embryos (see Stern & Downs, 2012 for review). Whether the yolk syncytial layer of teleost fish, or perhaps the yolky vegetal cells of amphibian embryos, constitute a functional equivalent in the context of the earliest stages of neural induction remains to be explored.

I would like to venture the proposal that the mechanisms of neural induction in amphibians and in the chick, if not all amniotes, are essentially the same, with only relatively minor differences in detail. Differences that appear major are probably due largely to differences in methodology dictated by the differences in topology and mechanical constraints on what can be done using conventional techniques in the various species.

References

Akai, J., & Storey, K. (2003). Brain or brawn: How FGF signaling gives us both. *Cell, 115*, 510–512.

Albazerchi, A., & Stern, C. D. (2007). A role for the hypoblast (AVE) in the initiation of neural induction, independent of its ability to position the primitive streak. *Developmental Biology, 301*, 489–503.

Alvarez, I. S., Araújo, M., & Nieto, M. A. (1998). Neural induction in whole chick embryo cultures by FGF. *Developmental Biology, 199*, 42–54.

Bachvarova, R. F., Skromne, I., & Stern, C. D. (1998). Induction of primitive streak and Hensen's node by the posterior marginal zone in the early chick embryo. *Development (Cambridge, England), 125*, 3521–3534.

Beddington, R. S. (1994). Induction of a second neural axis by the mouse node. *Development (Cambridge, England), 120*, 613–620.

Bellairs, R. (1986). The primitive streak. *Anatomy and Embryology, 174*, 1–14.

Bertocchini, F., & Stern, C. D. (2002). The hypoblast of the chick embryo positions the primitive streak by antagonizing nodal signaling. *Developmental Cell, 3*, 735–744.

Callebaut, M., & Van Nueten, E. (1994). Rauber's (Koller's) sickle: The early gastrulation organizer of the avian blastoderm. *European Journal of Morphology, 32*, 35–48.

Cambray, N., & Wilson, V. (2002). Axial progenitors with extensive potency are localised to the mouse chordoneural hinge. *Development (Cambridge, England), 129*, 4855–4866.

Camus, A., & Tam, P. P. (1999). The organizer of the gastrulating mouse embryo. *Current Topics in Developmental Biology, 45*, 117–153.

De Almeida, I., Rolo, A., Batut, J., Hill, C., Stern, C. D., & Linker, C. (2008). Unexpected activities of Smad7 in Xenopus mesodermal and neural induction. *Mechanisms of Development, 125*, 421–431.

Delaune, E., Lemaire, P., & Kodjabachian, L. (2005). Neural induction in Xenopus requires early FGF signalling in addition to BMP inhibition. *Development (Cambridge, England), 132*, 299–310.

Dias, A. S., De Almeida, I., Belmonte, J. M., Glazier, J. A., & Stern, C. D. (2014). Somites without a clock. *Science (New York, N. Y.), 343*, 791–795.

Dias, M. S., & Schoenwolf, G. C. (1990). Formation of ectopic neurepithelium in chick blastoderms: Age-related capacities for induction and self-differentiation following transplantation of quail Hensen's nodes. *The Anatomical Record, 228*, 437–448.

Downs, K. M., & Davies, T. (1993). Staging of gastrulating mouse embryos by morphological landmarks in the dissecting microscope. *Development (Cambridge, England), 118*, 1255–1266.

Eyal-Giladi, H., & Kochav, S. (1976). From cleavage to primitive streak formation: A complementary normal table and a new look at the first stages of the development of the chick I. General morphology. *Developmental Biology, 49*, 321–337.

Foley, A. C., Skromne, I., & Stern, C. D. (2000). Reconciling different models of forebrain induction and patterning: A dual role for the hypoblast. *Development (Cambridge, England), 127*, 3839–3854.

Foley, A. C., Storey, K. G., & Stern, C. D. (1997). The prechordal region lacks neural inducing ability, but can confer anterior character to more posterior neuroepithelium. *Development (Cambridge, England), 124*, 2983–2996.

Gallera, J. (1965). Quelle est la durée nécessaire pour déclencher des inductions neurales chez le poulet? *Experientia, 21*, 218–219.
Gallera, J. (1971a). Différence de la reactivité à l'inducteur neurogène entre l'ectoblaste de l'aire opaque et celui de l'aire pellucide chez le poulet. *Experientia, 26*, 1953–1954.
Gallera, J. (1971b). Primary induction in birds. *Advances in Morphogenesis, 9*, 149–180.
Gallera, J., & Ivanov, I. (1964). La competence neurogène du feuillet éxterne du blastoderme de poulet en fonction du facteur temps. *Journal of Embryology and Experimental Morphology, 12*, 693–711.
Gibson, A., Robinson, N., Streit, A., Sheng, G., & Stern, C. D. (2011). Regulation of programmed cell death during neural induction in the chick embryo. *The International Journal of Developmental Biology, 55*, 33–43.
Gouti, M., Tsakiridis, A., Wymeersch, F. J., Huang, Y., Kleinjung, J., Wilson, V., & Briscoe, J. (2014). In vitro generation of neuromesodermal progenitors reveals distinct roles for wnt signalling in the specification of spinal cord and paraxial mesoderm identity. *PLoS Biology, 12*, e1001937.
Gurdon, J. B. (1987). Embryonic induction—Molecular prospects. *Development (Cambridge, England), 99*, 285–306.
Hamburger, V., & Hamilton, H. L. (1951). A series of normal stages in the development of the chick embryo. *Journal of Morphology, 88*, 49–92.
Harland, R., & Gerhart, J. (1997). Formation and function of Spemann's organizer. *Annual Review of Cell and Developmental Biology, 13*, 611–667.
Hatada, Y., & Stern, C. D. (1994). A fate map of the epiblast of the early chick embryo. *Development (Cambridge, England), 120*, 2879–2889.
Hemmati-Brivanlou, A., & Melton, D. (1997). Vertebrate embryonic cells will become nerve cells unless told otherwise. *Cell, 88*, 13–17.
Hemmati-Brivanlou, A., & Thomsen, G. H. (1995). Ventral mesodermal patterning in Xenopus embryos: Expression patterns and activities of BMP-2 and BMP-4. *Developmental Genetics, 17*, 78–89.
Hensen, V. (1876). Beobachtungen über die Befruchtung und Entwicklung des Kaninchens und Meerschweinchens. *Zeitschr. f. Anatomie und Entwicklungsgesetz, 1*, 353–423.
Hopf, C., Viebahn, C., & Puschel, B. (2011). BMP signals and the transcriptional repressor BLIMP1 during germline segregation in the mammalian embryo. *Development Genes and Evolution, 221*, 209–223.
Izpisúa-Belmonte, J. C., De Robertis, E. M., Storey, K. G., & Stern, C. D. (1993). The homeobox gene goosecoid and the origin of organizer cells in the early chick blastoderm. *Cell, 74*, 645–659.
Joubin, K., & Stern, C. D. (1999). Molecular interactions continuously define the organizer during the cell movements of gastrulation. *Cell, 98*, 559–571.
Kinder, S. J., Tsang, T. E., Wakamiya, M., Sasaki, H., Behringer, R. R., Nagy, A., & Tam, P. P. (2001). The organizer of the mouse gastrula is composed of a dynamic population of progenitor cells for the axial mesoderm. *Development (Cambridge, England), 128*, 3623–3634.
Knoetgen, H., Teichmann, U., Wittler, L., Viebahn, C., & Kessel, M. (2000). Anterior neural induction by nodes from rabbits and mice. *Developmental Biology, 225*, 370–380.
Koller, C. (1882). Untersuchungen über die Blätterbildung im Hühnerkeim. *Archiv für mikroskopische Anatomie, 20*, 174–211.
Lamb, T. M., Knecht, A. K., Smith, W. C., Stachel, S. E., Economides, A. N., Stahl, N., ... Harland, R. M. (1993). Neural induction by the secreted polypeptide noggin. *Science (New York, N. Y.), 262*, 713–718.
Launay, C., Fromentoux, V., Shi, D. L., & Boucaut, J. C. (1996). A truncated FGF receptor blocks neural induction by endogenous Xenopus inducers. *Development (Cambridge, England), 122*, 869–880.

Lavial, F., Acloque, H., Bertocchini, F., Macleod, D. J., Boast, S., Bachelard, E., ... Pain, B. (2007). The Oct4 homologue PouV and Nanog regulate pluripotency in chicken embryonic stem cells. *Development (Cambridge, England), 134*, 3549–3563.

Linker, C., De Almeida, I., Papanayotou, C., Stower, M., Sabado, V., Ghorani, E., ... Stern, C. D. (2009). Cell communication with the neural plate is required for induction of neural markers by BMP inhibition: Evidence for homeogenetic induction and implications for Xenopus animal cap and chick explant assays. *Developmental Biology, 327*, 478–486.

Linker, C., & Stern, C. D. (2004). Neural induction requires BMP inhibition only as a late step, and involves signals other than FGF and Wnt antagonists. *Development (Cambridge, England), 131*, 5671–5681.

Mangold, O. (1933). Über die Induktionsfähighkeit der verschiedenen Bezirke der Neurula von Urodelen. *Naturwissenshaften, 21*, 761–766.

McGrew, M. J., Sherman, A., Lillico, S. G., Ellard, F. M., Radcliffe, P. A., Gilhooley, H. J., ... Sang, H. (2008). Localised axial progenitor cell populations in the avian tail bud are not committed to a posterior Hox identity. *Development (Cambridge, England), 135*, 2289–2299.

Metzis, V., Steinhauser, S., Pakanavicius, E., Gouti, M., Stamataki, D., Ivanovitch, K., ... Briscoe, J. (2018). Nervous system regionalization entails axial allocation before neural differentiation. *Cell, 175*, 1105–1118 e1117.

Multi-author. (1992). Postimplantation development in the mouse. In D. J. Chadwick & J. Marsh (Eds.), *Symposium*. London, 3–5 June 1991. Ciba Foundation Symposium 165, pp. 1–303.

Nicolet, G. (1968). Le rôle du noeud de Hensen sur la différenciation des somites chez le poulet. *Experientia, 24*, 263–264.

Nicolet, G. (1970). Is the presumptive notochord responsible for somite genesis in the chick? *Journal of Embryology and Experimental Morphology, 24*, 467–478.

Nicolet, G. (1971). The young notochord can induce somite genesis by means of diffusible substances in the chick. *Experientia, 27*, 938–939.

Nieuwkoop, P. D. (1969a). The formation of mesoderm in urodelean amphibians I. Induction by the endoderm. *Wilhelm Roux Arch Entwicklungsmech. Organismen, 162*, 341–373.

Nieuwkoop, P. D. (1969b). The formation of mesoderm in Urodelean amphibians. II. The origin of the dorso-ventral polarity of the mesoderm. *Wilhelm Roux' Arch Entwicklungsmech. Organismen, 163*, 298–315.

Nieuwkoop, P. D., Botternenbrood, E. C., Kremer, A., Bloesma, F. F. S. N., Hoessels, E. L. M. J., Meyer, G., & Verheyen, F. J. (1952). Activation and organization of the central nervous system in amphibians. *The Journal of Experimental Zoology, 120*, 1–108.

Papanayotou, C., De Almeida, I., Liao, P., Oliveira, N. M. M., Lu, S.-Q., Kougioumtzidou, E., ... Stern, C. D. (2013). Calfacilitin is a calcium channel modulator essential for initiation of neural plate development. *Nature Communications, 4*, 1837.

Papanayotou, C., Mey, A., Birot, A. M., Saka, Y., Boast, S., Smith, J. C., ... Stern, C. D. (2008). A mechanism regulating the onset of Sox2 expression in the embryonic neural plate. *PLoS Biology, 6*, e2.

Pera, E., Ikeda, A., Eivers, E., & De Robertis, E. M. (2003). Integration of IGF, FGF and anti-BMP signals via Smad1 phosphorylation in neural induction. *Genes & Development, 17*, 3023–3028.

Perea-Gómez, A., Vella, F. D., Shawlot, W., Oulad-Abdelghani, M., Chazaud, C., Meno, C., ... Ang, S. L. (2002). Nodal antagonists in the anterior visceral endoderm prevent the formation of multiple primitive streaks. *Developmental Cell, 3*, 745–756.

Piccolo, S., Sasai, Y., Lu, B., & De Robertis, E. M. (1996). Dorsoventral patterning in xenopus: Inhibition of ventral signals by direct binding of chordin to BMP-4. *Cell, 86*, 589–598.

Pinho, S., Simonsson, P. R., Trevers, K. E., Stower, M. J., Sherlock, W. T., Khan, M., ... Stern, C. D. (2011). Distinct steps of neural induction revealed by Asterix, Obelix and TrkC, genes induced by different signals from the organizer. *PLoS One, 6*, e19157.

Psychoyos, D., & Stern, C. D. (1996). Fates and migratory routes of primitive streak cells in the chick embryo. *Development (Cambridge, England), 122*, 1523–1534.

Robb, L., & Tam, P. P. (2004). Gastrula organiser and embryonic patterning in the mouse. *Seminars in Cell & Developmental Biology, 15*, 543–554.

Rodríguez-Gallardo, L., Climent, V., Garcia-Martinez, V., Schoenwolf, G. C., & Alvarez, I. S. (1997). Targeted over-expression of FGF in chick embryos induces formation of ectopic neural cells. *The International Journal of Developmental Biology, 41*, 715–723.

Rosenquist, G. C. (1966). A Radioautographic study of labelled grafts in the chick blastoderm. Development from primitive-streak stages to stage 12. *Contributions to Embryology Carnegie Institution of Washington, 38*, 71–110.

Rowan, A. M., Stern, C. D., & Storey, K. G. (1999). Axial mesendoderm refines rostrocaudal pattern in the chick nervous system. *Development (Cambridge, England), 126*, 2921–2934.

Sanchez-Arrones, L., Ferran, J. L., Rodriguez-Gallardo, L., & Puelles, L. (2009). Incipient forebrain boundaries traced by differential gene expression and fate mapping in the chick neural plate. *Developmental Biology, 335*, 43–65.

Sanchez-Arrones, L., Stern, C. D., Bovolenta, P., & Puelles, L. (2012). Sharpening of the anterior neural border in the chick by rostral endoderm signalling. *Development (Cambridge, England), 139*, 1034–1044.

Sasai, Y., Lu, B., Piccolo, S., & De Robertis, E. M. (1996). Endoderm induction by the organizer-secreted factors chordin and noggin in Xenopus animal caps. *The EMBO Journal, 15*, 4547–4555.

Sasai, Y., Lu, B., Steinbeisser, H., & De Robertis, E. M. (1995). Regulation of neural induction by the Chd and Bmp-4 antagonistic patterning signals in Xenopus. *Nature, 376*, 333–336.

Sasai, Y., Lu, B., Steinbeisser, H., Geissert, D., Gont, L. K., & De Robertis, E. M. (1994). Xenopus chordin: A novel dorsalizing factor activated by organizer-specific homeobox genes. *Cell, 79*, 779–790.

Saxen, L. (1980). Neural induction: Past, present, and future. *Current Topics in Developmental Biology, 15*(Pt 1), 409–418.

Schoenwolf, G. C. (1992). Morphological and mapping studies of the paranodal and postnodal levels of the neural plate during chick neurulation. *The Anatomical Record, 233*, 281–290.

Schoenwolf, G. C., Bortier, H., & Vakaet, L. (1989). Fate mapping the avian neural plate with quail/chick chimeras: Origin of prospective median wedge cells. *The Journal of Experimental Zoology, 249*, 271–278.

Schoenwolf, G. C., & Sheard, P. (1990). Fate mapping the avian epiblast with focal injections of a fluorescent- histochemical marker: Ectodermal derivatives. *The Journal of Experimental Zoology, 255*, 323–339.

Selleck, M. A., & Stern, C. D. (1991). Fate mapping and cell lineage analysis of Hensen's node in the chick embryo. *Development (Cambridge, England), 112*, 615–626.

Selleck, M. A. J., & Stern, C. D. (1992). Evidence for stem cells in the mesoderm of Hensen's node and their role in embryonic pattern formation. In R. Bellairs, E. J. Sanders, & J. W. Lash (Eds.). *Formation and differentiation of early embryonic mesoderm* (pp. 23–31). New York: Plenum Press.

Sheng, G., Dos Reis, M., & Stern, C. D. (2003). Churchill, a zinc finger transcriptional activator, regulates the transition between gastrulation and neurulation. *Cell, 115*, 603–613.

Slack, J. (1991). *From egg to embryo: Regional specification in early development* (2nd ed.). Cambridge: Cambridge University Press.

Slack, J. (2024). The organizer: What it meant, and still means, to developmental biology. *Curr. Top. Dev. Biol. 157*, 1–42.

Solovieva, T., Lu, H. C., Moverley, A., Plachta, N., & Stern, C. D. (2022a). The embryonic node behaves as an instructive stem cell niche for axial elongation. *Proceedings of the National Academy of Sciences of the United States of America, 119*, e2108935119.

Solovieva, T., Wilson, V., & Stern, C. D. (2022b). A niche for axial stem cells—A cellular perspective in amniotes. *Developmental Biology, 490*, 13–21.

Spemann, H. (1921). Die Erzeugung thierischer Chimären durch heteroplastische Transplantation zwischen Triton cristatus und taeniatus. *Wilhelm Roux' Arch Entwicklungsmech. Organismen, 48*, 533–570.

Spemann, H., & Mangold, H. (1924). Über Induktion von Embryonalanlagen durch Implantations artfremder Organisatoren. *Roux's Arch Entwicklungsmech. Organismen, 100*, 599–638.

Stern, C. D. (2005). Neural induction: Old problem, new findings, yet more questions. *Development (Cambridge, England), 132*, 2007–2021.

Stern, C. D., & Downs, K. M. (2012). The hypoblast (visceral endoderm): An evo-devo perspective. *Development (Cambridge, England), 139*, 1059–1069.

Storey, K. G., Crossley, J. M., De Robertis, E. M., Norris, W. E., & Stern, C. D. (1992). Neural induction and regionalisation in the chick embryo. *Development (Cambridge, England), 114*, 729–741.

Storey, K. G., Goriely, A., Sargent, C. M., Brown, J. M., Burns, H. D., Abud, H. M., & Heath, J. K. (1998). Early posterior neural tissue is induced by FGF in the chick embryo. *Development (Cambridge, England), 125*, 473–484.

Streit, A., Berliner, A. J., Papanayotou, C., Sirulnik, A., & Stern, C. D. (2000). Initiation of neural induction by FGF signalling before gastrulation. *Nature, 406*, 74–78.

Streit, A., Lee, K. J., Woo, I., Roberts, C., Jessell, T. M., & Stern, C. D. (1998). Chordin regulates primitive streak development and the stability of induced neural cells, but is not sufficient for neural induction in the chick embryo. *Development (Cambridge, England), 125*, 507–519.

Streit, A., Sockanathan, S., Pérez, L., Rex, M., Scotting, P. J., Sharpe, P. T., ... Stern, C. D. (1997). Preventing the loss of competence for neural induction: HGF/SF, L5 and Sox-2. *Development (Cambridge, England), 124*, 1191–1202.

Streit, A., & Stern, C. D. (1999a). Establishment and maintenance of the border of the neural plate in the chick: Involvement of FGF and BMP activity. *Mechanisms of Development, 82*, 51–66.

Streit, A., & Stern, C. D. (1999b). Mesoderm patterning and somite formation during node regression: Differential effects of chordin and noggin. *Mechanisms of Development, 85*, 85–96.

Tam, P. P., & Steiner, K. A. (1999). Anterior patterning by synergistic activity of the early gastrula organizer and the anterior germ layer tissues of the mouse embryo. *Development (Cambridge, England), 126*, 5171–5179.

Thiery, A., Buzzi, A. L., & Streit, A. (2020). Cell fate decisions during the development of the peripheral nervous system in the vertebrate head. *Current Topics in Developmental Biology, 139*, 127–167.

Thiery, A. P., Buzzi, A. L., Hamrud, E., Cheshire, C., Luscombe, N. M., Briscoe, J., & Streit, A. (2023). scRNA-sequencing in chick suggests a probabilistic model for cell fate allocation at the neural plate border. *Elife, 12*, e82717.

Thomas, P., & Beddington, R. (1996). Anterior primitive endoderm may be responsible for patterning the anterior neural plate in the mouse embryo. *Current Biology: CB, 6*, 1487–1496.

Tonegawa, A., Funayama, N., Ueno, N., & Takahashi, Y. (1997). Mesodermal subdivision along the mediolateral axis in chicken controlled by different concentrations of BMP-4. *Development (Cambridge, England), 124*, 1975–1984.

Tonegawa, A., & Takahashi, Y. (1998). Somitogenesis controlled by Noggin. *Developmental Biology, 202*, 172–182.

Trevers, K. E., Lu, H. C., Yang, Y., Thiery, A. P., Strobl, A. C., Anderson, C., ... Stern, C. D. (2023). A gene regulatory network for neural induction. *Elife, 12*, e73189.

Trevers, K. E., Prajapati, R. S., Hintze, M., Stower, M. J., Strobl, A. C., Tambalo, M., ... Streit, A. (2018). Neural induction by the node and placode induction by head mesoderm share an initial state resembling neural plate border and ES cells. *Proceedings of the National Academy of Sciences of the United States of America, 115*, 355–360.

Uchikawa, M., Ishida, Y., Takemoto, T., Kamachi, Y., & Kondoh, H. (2003). Functional analysis of chicken Sox2 enhancers highlights an array of diverse regulatory elements that are conserved in mammals. *Developmental Cell, 4*, 509–519.

Uchikawa, M., Yoshida, M., Iwafuchi-Doi, M., Matsuda, K., Ishida, Y., Takemoto, T., & Kondoh, H. (2011). B1 and B2 Sox gene expression during neural plate development in chicken and mouse embryos: Universal versus species-dependent features. *Development, Growth & Differentiation, 53*, 761–771.

Voiculescu, O., Bertocchini, F., Wolpert, L., Keller, R. E., & Stern, C. D. (2007). The amniote primitive streak is defined by epithelial cell intercalation before gastrulation. *Nature, 449*, 1049–1052.

Voiculescu, O., Bodenstein, L., Lau, I. J., & Stern, C. D. (2014). Local cell interactions and self-amplifying individual cell ingression drive amniote gastrulation. *Elife, 3* e01817.

Waddington, C. H. (1933). Induction by the primitive streak and its derivatives in the chick. *Journal of Experimental Biology, 10*, 38–48.

Waddington, C. H. (1934). Experiments on embryonic induction. *The Journal of Experimental Biology, 11*, 211–227.

Waddington, C. H. (1936). Organizers in mammalian development. *Nature, 138*, 125.

Waddington, C. H. (1937). Experiments on determination in the rabbit embryo. *Archives of Biology, 48*, 273–290.

Waddington, C. H., & Schmidt, G. A. (1933). Induction by heteroplastic grafts of the primitive streak in birds. *Wilhelm Roux Arch Entwicklungsmech. Organismen, 128*, 522–563.

Waddington, C. H., & Waterman, A. J. (1933). The development in vitro of young rabbit embryos. *Journal of Anatomy, 67*, 355–370.

Wilson, P. A., & Hemmati-Brivanlou, A. (1995). Induction of epidermis and inhibition of neural fate by Bmp-4. *Nature, 376*, 331–333.

Wilson, S. I., Rydstrom, A., Trimborn, T., Willert, K., Nusse, R., Jessell, T. M., & Edlund, T. (2001). The status of Wnt signalling regulates neural and epidermal fates in the chick embryo. *Nature, 411*, 325–330.

Zimmerman, L. B., De Jesus-Escobar, J. M., & Harland, R. M. (1996). The Spemann organizer signal noggin binds and inactivates bone morphogenetic protein 4. *Cell, 86*, 599–606.

CHAPTER THREE

Tissues and signals with true organizer properties in craniofacial development

Shruti S. Tophkhane and Joy M. Richman*

Life Sciences Institute and Faculty of Dentistry, University of British Columbia, Vancouver, BC, Canada
*Corresponding author: e-mail address: richman@dentistry.ubc.ca

Contents

1. Introduction	68
2. Neural crest cells contain craniofacial patterning information, but do they have organizer properties?	69
3. Foregut endoderm is an organizer in facial patterning	69
4. Nasal placodes as craniofacial organizers	74
5. Frontonasal mass epithelial zone a potential facial organizer	75
6. Signals with craniofacial organizer properties	75
6.1 Sonic Hedgehog and Fibroblast growth factor	75
6.2 Noggin and retinoic acid	77
6.3 Endothelin	78
7. Concluding remarks	79
Acknowledgments	79
References	80

Abstract

Transplantation experiments have shown that a true organizer provides instructive signals that induce and pattern ectopic structures in the responding tissue. Here, we review craniofacial experiments to identify tissues with organizer properties and signals with organizer properties. In particular, we evaluate whether transformation of identity took place in the mesenchyme. Using these stringent criteria, we find the strongest evidence for the avian foregut ectoderm. Transplanting a piece of quail foregut endoderm to a host chicken embryo caused ectopic beaks to form derived from chicken mesenchyme. The beak identity, whether upper or lower as well as orientation, was controlled by the original anterior-posterior position of the donor endoderm. There is also good evidence that the nasal pit is necessary and sufficient for lateral nasal patterning. Finally, we review signals that have organizer properties on their own without the need for tissue transplants. Mouse germline knockouts of the endothelin pathway result in transformation of identity of the mandible into a maxilla.

Application of noggin-soaked beads to post-migratory neural crest cells transforms maxillary identity. This suggests that endothelin or noggin rich ectoderm could be organizers (not tested). In conclusion, craniofacial, neural crest-derived mesenchyme is competent to respond to tissues with organizer properties, also originating in the head. In future, we can exploit such well defined systems to dissect the molecular changes that ultimately lead to patterning of the upper and lower jaw.

1. Introduction

An organizer can be defined as a group of cells derived from any germ layer that recruits neighboring cells and induces a change in the patterning and fate of host cells (Gurdon, 1987; Waddington, 1966). The best proof of a tissue possessing organizer properties comes from grafting experiments. The grafting and ablation experiments have shown that if the donor graft is an organizer, the graft will recruit surrounding host cells to participate in forming ectopic structures. The classic example of an organizer was shown by transplantation of the dorsal lip of the blastopore of an unpigmented newt into ventral regions of a pigmented newt embryo as reviewed by Anderson and Stern (2016). The unpigmented donor cells induced the host cells to form a complete secondary axis. Two possible outcomes exist for these and other organizer grafts. As defined, the patterning of the ectopic structure should be dictated by the source of the organizer. The ectopic structure should not branch from the endogenous structure. As we will see, in the face there are a subset of grafting experiments in which the grafts do not change the patterning of the host mesenchyme. Instead, the graft induces duplicated structures beside the normal structures and identity is not reprogrammed. In all grafting experiments, it is crucial to rule out tissue contamination where some of the donor mesenchyme is transferred along with donor epithelium. It is also necessary to determine whether distant host cells with the correct patterning information migrate towards the grafted tissue and then form the ectopic structures. Finally, the ultimate proof is that signals from organizers can be delivered as proteins or genes and these can accomplish some or all the de novo patterning in situ without any grafting of tissues or migration of host cells. In this review we will focus on craniofacial organizers which were identified principally by grafting experiments in avian embryos. As we will see, unlike other parts of the body, all the organizers in the face are epithelial (endodermal or ectodermal) derived.

2. Neural crest cells contain craniofacial patterning information, but do they have organizer properties?

Craniofacial development begins with the initiation of pluripotent migratory neural crest cells at the border between the neural and non-neural ectoderm (Candido-Ferreira et al., 2023). Quail-chicken chimeras were the first to show that neural crest cells have patterning information required to form the facial skeleton (Couly et al., 1993; Le Lievre, 1978; Noden, 1978) and non-skeletal derivatives (Etchevers et al., 2001; Koentges & Lumsden, 1996; Noden & Francis-West, 2006). The facial skeleton is derived from *HOX*-negative neural crest cells adjacent to the diencephalic, mesencephalic, and anterior rhombencephalic neural folds (Couly et al., 1998). All *HOX*-negative neural crest cells have equivalent ability to form facial bones as shown in orthotopic quail-chicken grafting experiments (Couly et al., 1998; Noden, 1978). There also some intrinsic patterning embedded in the *HOX*-negative neural crest cells as shown in heterotopic grafts where neural folds from the mesencephalon down to r2 were grafted to r4 and formed ectopic mandibular skeleton in the neck (Couly et al., 1998; Noden, 1983). The grafts continued to be *HOXA2*-negative and suppressed host *HOXA2* expression in the donor site (Couly et al., 1998) which was necessary for development of the ectopic mandibular skeleton. Subsequently, the exact location of neural crest cells that contained the intrinsic patterning for the mandible was identified. Neural tube grafts consisting of just the mid-hind brain isthmus and premigratory neural crest cells could also induce ectopic mandibular structures in the second arch. Mouse knockouts of *Hoxa2* reveal the importance of this gene for second pharyngeal arch development. In the complete absence of *Hoxa2* several first arch derivatives form in the second arch (Gendron-Maguire et al., 1993; Rijli et al., 1993). Thus *Hoxa2* in chicken and mouse is necessary and sufficient to regulate second pharyngeal arch identity. However, are neural crest cells acting as organizers in these heterotopic grafting experiments? Since donor neural crest cells from quail make all of the ectopic structures, there is no evidence that the host chicken embryo is responding to these cues (Table 1).

3. Foregut endoderm is an organizer in facial patterning

The foregut grafting experiments in quail and chicken embryos confirmed that the endoderm in the head qualifies as organizer for

Table 1 List of potential craniofacial regions and signals with organizer properties.

	Proof that epithelial organizer does not contribute to ectopic mesenchymal structures (YES)	Ectopic structure must be derived from host mesenchyme (YES)	Ectopic structure is a branch of the normal element (YES or NO)	Transformation of identity is induced (YES or NO)	Signals identified in the organizer and verified in vivo (YES or NO)
CNCC (Couly et al., 1998; Noden, 1983)	No	No, ectopic cartilage in neck is quail-derived	No	No, novel structure is donor derived, no change in host embryo pattern	No
Mesencephalic level foregut endoderm (Couly et al., 2002)	Yes, quail cells were only in the epithelium	Yes, chicken cells made up the ectopic beak skeleton	No, separate supernumerary mandibles formed	No, extra lower beak originated from 1st pharyngeal arch so original patterning is maintained	Yes, SHH-secreting donor fibroblasts recreate the branched PA1 and mandible
Prosencephalic level foregut endoderm (Benouaiche et al., 2008)	Yes, yes quail cells were only in the epithelium	Yes, chicken cells made up ectopic beak skeleton	No	Yes, lower beak transformed to upper beak midline	Yes, in vivo with SHH bead implants and in vitro with NCC and endoderm co-cultures

FEZ (Hu & Helms, 1999; Hu & Marcucio, 2009; Hu, Marcucio, & Helms, 2003)	Yes	Yes	yes, branch of the normal prenasal cartilage	No, original identity maintained	Yes, FGF8, SHH
Nasal pit (Szabo-Rogers et al., 2009)	Yes, quail cells found only in nasal pit epithelium not in mesenchyme	Yes, only chicken cells contributed to ectopic bone	No	Uncertain but mesenchyme identity in maxilla is transformed to lateral nasal shown by *PAX7* expression	Yes, FGF required but not sufficient to induce nasal passages
Noggin (Celá et al., 2016; Lee et al., 2001; Nimmagadda et al., 2015)	N/A	N/A	No	Yes, Noggin beads changed identity of maxillary prominence as shown by induction of egg tooth	Yes, Noggin is the main signal
Endothelin pathway (Kurihara et al., 1994; Ozeki et al., 2004; Ruest et al., 2004; Sato et al., 2008)	N/A	N/A	No	Yes, mandible converted to maxilla in *Edn1−/−* embryos. *Edn1* is expressed in maxillary prominence changes maxillary to mandibular	Yes, *Edn1* is mediated by *Dlx5/6*

craniofacial patterning. When foregut endoderm from quail was grafted into the head of stage-matched chick embryo, the graft retained the positional information and directed chicken mesenchyme to form supernumerary or duplicated lower beaks alongside the normal lower beak (Couly et al., 2002) (stripe II; Fig. 1A). The authors provided limited histological analysis of embryos at older stages to show that the source of the ectopic bones was chicken. Adherent mesodermal cells may have been transplanted along with the endoderm. However, mesoderm does not give rise to the skeleton (Couly et al., 1992) and in the limited histology from the chimeras with endoderm there were no quail cells were in the muscle (Couly et al., 2002). Another possible explanation for the ectopic beaks is that the foregut causes a change in the migration patterns of chicken neural crest cells. The foregut grafts are placed prior to the onset of migration of neural crest cells and it is possible that some of the mandibular neural crest cells are induced to migrate caudally into the second arch.

Moving further anterior in the head, the anterior piece of endoderm was determined to be necessary and sufficient for frontonasal patterning. Ablation of the anterior endoderm in chicken embryo led to absence of mesethmoid (the cartilage primordium of the upper beak in chicken) (Benouaiche et al., 2008; Couly et al., 2002). Grafting experiments of anterior foregut endoderm which expresses SHH led to formation of supernumerary mesethmoid –like cartilage present adjacent to the normal nasal capsule (Benouaiche et al., 2008). When the same graft was placed in the presumptive 2nd pharyngeal arch territory, supernumerary mesethmoid-like element formed below the mandible (Benouaiche et al., 2008) (Stripe 1, Fig. 1A). Thus, the anterior foregut endoderm was able to recruit the mesenchyme from the chicken that would normally have formed the hyoid skeleton to form skeletal elements derived from the forebrain neural crest cells (Couly et al., 1992). Again the caveat is that cranial neural crest cells are a highly migratory population so part of the phenotypes could be attributed to ectopic migration of cells fated to make one jaw region into another. However, based on the lack of patterning changes seen when grafts of neural crest cells are transplanted within the HOXA2-negative region, the mismigration alone is insufficient to explain the results. Overall, the authors from the Le Douarin lab were the first to confirm that the posterior foregut endoderm harbors instructive signals that can direct the HOX-negative cranial neural crest cells to form facial skeleton.

Tissues and signals with true organizer properties in craniofacial development 73

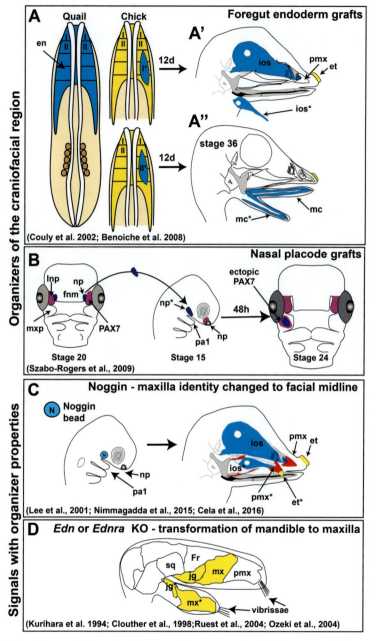

Fig. 1 Organizers and signals with organizing properties. (A) Foregut endoderm graft from quail to chicken embryos at 5 somite stage. (A′) Stripe I endoderm graft placed caudally duplicated the frontonasal mass cartilage elements in the hyoid region (Benoiche et al., 2008). (A″) Stripe II endoderm graft duplicated the lower jaw in the
(Continued)

The authors of these original studies never used the term "facial organizer"; however, we propose that foregut endoderm is a true craniofacial organizer (Table 1).

4. Nasal placodes as craniofacial organizers

Sensory placodes in the head do have important roles in patterning the craniofacial ganglia (Webb & Noden, 1993), however the role in skeletal patterning of the optic, olfactory and otic placodes have only rarely been studied. Our lab performed ablation and grafting experiments of nasal placodes after neural crest cell migration was completed, to investigate if the placode is required and sufficient to pattern the nearby facial skeleton (Szabo-Rogers et al., 2008). Ablation of the nasal placode led to missing nasal bone and nasal conchae together with loss of expression of *FGF8*, *DLX5*, and *PAX7*. Adding back a bead soaked in FGF8 restored the nasal bone. Grafting of quail nasal placodes (stage 20) into chicken maxillary prominence (an area distant from the lateral nasal prominence) (stage 15) led to partial transformation of maxilla to lateral nasal identity including expression of paired box 7 (*PAX7*) which is localized to only lateral nasal mesenchyme (Fig. 1B). Histological analysis confirmed that the supernumerary skeletal structures derived from the chicken, while the olfactory neurons were derived from quail. Others have also demonstrated the fate of the craniofacial placodes in neurogenesis and ganglion formation (Bhattacharyya & Bronner-Fraser, 2008; Bhattacharyya et al., 2004). Although identity of the ectopic bones was

Fig. 1—Cont'd hyoid region (Couly et al., 2002). (B) Nasal pit epithelium graft from stage 20 quail donor embryos grafted into the post-optic region of stage 15 chicken host. Expression of PAX7 protein (pink) is normally restricted to the lateral nasal process. After graft has grown for 48 h, ectopic PAX7 is induced in the maxillary prominence around the graft (pink). After 10 days, skeletal elements form on the side of the maxilla (not shown). (C) When BMP signaling is blocked by placing a bead soaked in noggin in the post optic region, maxillary identity is lost and replaced by frontonasal derivatives including premaxilla, interorbital septum and egg tooth were observed. (D) Germline knockouts of *Edn* or *Ednra* caused replacement of mandibular skeletal elements with maxillary along with ectopic vibrissae on the lower jaw. Key: * – supernumerary structures or grafts, en – endoderm, et – egg tooth, fnm – frontonasal mass, fr – frontal bone, ios – interorbital septum, jg – jugal bone, lnp – lateral nasal prominence, mc – Meckel's cartilage; mx – maxillary bone, mxp – maxillary prominence, np – nasal pit, pa1 – pharyngeal arch 1, pmx – premaxilla, sq – squamosal bone.

not clear, these structures originated from the chicken host hence cells were responding to the grafted nasal placode. Based on loss- and gain-of-function experiments, the nasal placodes and FGF8 have craniofacial organizer properties (Table 1).

5. Frontonasal mass epithelial zone a potential facial organizer

The Frontonasal Ectodermal Zone (FEZ)(Hu et al., 2003; Hu and Marcucio, 2009) is a region of facial ectoderm extending from the frontonasal mass into the stomodeum that expresses Fibroblast growth factor 8 (*FGF8*)(Hu et al., 2003), Sonic Hedgehog (*SHH*) (Hu et al., 2003) and other genes such as *BMP4* and *NOGGIN* (Ashique et al., 2002). A full profile of gene expression of the FEZ has not yet been carried out. The FEZ has been proposed to be an organizer of facial region and required for normal patterning and growth of the frontonasal prominence (Hu et al., 2003). Heterochronic, FEZ grafted from stage 20 quail to stage 25 chicken, the upper beak was bifurcated and had additional egg tooth (Hu et al., 2003). Transplantation of an equivalent mouse FEZ to chicken did not cause the chicken to resemble a mouse, but instead the mouse FEZ instructed the chicken mesenchyme to form a branched prenasal cartilage (Hu and Marcucio, 2009). Furthermore, when the FEZ was transplanted into the mandibular prominence, Meckel's cartilage bifurcated (twinned mandible) (Hu et al., 2003). The identity of the cartilage was still mandibular, so the graft did not carry positional information. Thus, the FEZ is sufficient to re-organize facial mesenchyme and epithelial patterning but is not a canonical organizer since the identity of the host mesenchyme is not altered. In addition, the requirement for the FEZ in developing upper beaks has not been demonstrated in extirpation experiments.

6. Signals with craniofacial organizer properties
6.1 Sonic Hedgehog and Fibroblast growth factor

SHH expression is restricted in the anterior part of foregut endoderm at 6ss in the chicken embryos (Brito et al., 2006). Ablation of foregut endoderm in chicken embryos (6ss) removed *Shh* expression resulting in massive apoptosis and absence of both upper and lower beaks. Excision of foregut endoderm at 8- to 10-ss leads to absence of upper beak but normal lower beak. Recombinant SHH beads (100 μg/mL) placed into the forebrain region at 6ss were partially able to rescue the cell death and majority of the beak phenotype (Brito et al., 2006). These experiments suggested that organizer properties of foregut endoderm to might be due to the SHH. Therefore, the authors went on to graft SHH-producing QT6 fibroblasts into host chicken embryos. Cells were grafted into the first pharyngeal arch, an area affected by foregut endoderm grafts (Couly et al., 2002). Interestingly, SHH fibroblasts do not contribute to the host skeleton but led to striking lower jaw triplication (Brito et al., 2008). The SHH fibroblasts caused mirror-image branching of the normal Meckel's cartilage. The earlier stages of the embryos confirmed the mirroring of the first arch in terms of gene expression and morphology. In contrast, endoderm grafts induced full lower jaw duplication in the hyoid region or outside the normal domain for the mandibular skeleton (Couly et al., 2002). Thus, SHH likely controls morphogenesis of Meckel's cartilage and may be one of several signals contributing to the organizer capacity of the foregut endoderm. An instructive role has been demonstrated in the frontonasal mass. Virus expressing cells grafted into the frontonasal mass [similar to (Brito et al., 2008)] infect the host chicken mesenchyme and led to branches of the prenasal cartilage or supernumerary prenasal cartilages capped with egg teeth (Hu & Helms, 1999). The skeletal phenotypes were not fully described so whether branching off the normal prenasal cartilage occurred still needs to be confirmed. SHH beads were implanted into the frontonasal mass at stage 25 but these beads did not change patterning (Hu & Helms, 1999). Therefore, when SHH is delivered by cells which secrete high levels of the protein over prolonged periods of time, facial patterning is reorganized. However, jaw identity is never altered in such experiments.

The authors of the original FEZ study investigated the role of FGF in facial patterning in more detail (Hu et al., 2003). Transplanted epithelium (both endodermal and ectodermal) from stage 25 donors that no longer expressed *FGF8* had no effect on patterning. Adding an FGF2-soaked bead to the graft supported a short branch of the prenasal cartilage and a second egg tooth (Hu et al., 2003). However, single beads soaked in FGF2 without a graft of the FEZ were insufficient to change facial patterning. Further studies using *FGF8*-expressing viruses showed that outgrowths containing cartilage were induced but these did not have patterning characteristic of a specific anatomical region (Abzhanov & Tabin, 2004). When *FGF8* and *SHH* viruses were combined, many outgrowths formed and those in the dorsal upper beak resembled the outgrowths induced by the FEZ (Abzhanov & Tabin, 2004). This suggests a synergy between FGF and SHH in organizing facial mesenchyme but even when expressed jointly, specific features of upper or lower jaws were not induced.

6.2 Noggin and retinoic acid

A series of experiments from our lab identified roles for BMP specifying maxillary and frontonasal mass identity. In these studies, no tissue grafts were performed. Instead, 2 beads were implanted into stage 15 embryos, after the cessation of neural crest migration. The original study showed that blocking BMP signaling with the BMP-antagonist Noggin was sufficient to change the identity of the maxillary prominence to frontonasal. The maxillary skeleton was replaced by prenasal cartilage, premaxilla and an egg tooth (Lee et al., 2001) (Fig. 1C). Adding a second bead soaked in RA increased the penetrance of the phenotype but all-trans retinoic acid on its own did not change patterning (Lee et al., 2001). We checked for migration of frontonasal cells into the maxillary region using DiI labeling and found no evidence of labeled cells moving out of the frontonasal mass. The molecular changes underlying the phenotypes induced by RA and Noggin were profiled with microarray analysis (Nimmagadda et al., 2015). Part of the reason retinoic acid synergized with noggin is that one of the targets of noggin is the RA synthesis enzyme, *ALDH1A3*. Novel targets of the retinoic acid and BMP pathways were also identified including the gene *PI15* (codes for Peptidase Inhibitor 15) (Nimmagadda et al., 2015). Expression of a retrovirus containing *PI15* combined with low-dose Noggin (too low to elicit a phenotype) was able to transform the maxilla into frontonasal mass (Nimmagadda et al., 2015). However on its own, overexpression of *PI15* induced formation of clefts in the upper beak (Nimmagadda et al., 2015). These clefts resemble the phenotypes induced by RA beads implanted into the face (Richman & Delgado, 1995). Thus, *PI15* may have been the target of RA that synergized with noggin. However, *PI15* on its own does not have organizer properties.

Further investigation into the spatial and temporal effects of noggin was carried out. Only beads soaked in noggin placed in specific regions of post-optic mesenchyme at stage were able to instruct transformation of identity of maxillary bone into a premaxilla. Beads placed closest to the oral epithelium in stage 15 chicken embryos were consistently able to induce duplication of cartilages, interorbital septum, and egg tooth along with the loss of maxillary bones (Celá et al., 2016). In contrast, implanting noggin beads at stage 20 inhibited bone formation in the maxilla (Celá et al., 2016).

The inhibition of ossification by noggin in stage 20 embryos was expected based on other studies (Ashique et al., 2002; Hu et al., 2008; Pizette & Niswander, 2000). There were stage specific, molecular responses of the post-migratory neural crest-derived mesenchyme that correlate with these phenotypes. For example, noggin beads led to strong upregulation of *SOX9* and decreased *RUNX2* expression compared to controls that had PBS beads implanted. The upregulation of a chondrogenesis transcription factor *SOX9* could set the stage for the ectopic cartilage that will form in the upper beak. Importantly, noggin beads placed directly into the first arch did not affect mandibular patterning. Therefore, maintaining relatively high BMP levels is specifically required to establish the identity of the maxillary prominence. The endogenous organizer is predicted to be a high *NOG*-expressing area of the frontonasal mass. Thus far we have reported *NOG* expression in a stripe of frontonasal mass ectoderm overlapping the FEZ at stages 24 and 28 (Ashique et al., 2002). There is limited data showing expression of *NOG* in the lateral stomodeal epithelium in the same location and stage as beads were implanted (Bothe et al., 2011). Thus it would be interesting to test whether such regions of BMP antagonist-expressing facial ectoderm have organizer properties.

6.3 Endothelin

G-protein coupled endothelin receptor-A (codes for *Ednra*) is expressed in the cranial neural crest derived ectomesenchyme (Clouthier et al., 2000). The primary ligand, endothelin-1 (codes for *Edn*) is expressed in the ectodermal epithelium of the pharyngeal arches and endodermal epithelium of pharyngeal pouch clefts (Clouthier et al., 1998). Endothelin-converting enzyme (codes for *Ece1*) is required for converting endothelin-1 into its active form and is expressed at sites where *Ednra* is expressed (Yanagisawa et al., 1998). The ligand, receptor and converting enzyme are all required to pattern mandibular identity.

Mice with a complete loss-of-function in *Edn* lose mandibular identity although at the time the authors did not recognize that maxillary identity had been gained (Kurihara et al., 1994). Later the *Edn* knockout mice were investigated in more detail. A second set of vibrissae appeared on the lower jaw of *Edn* null embryos (normally restricted to upper jaw) and maxillary skeleton replaced key mandibular elements (Fig. 1D) (Ozeki et al., 2004). Subsequent knockouts of *Endra* (Clouthier et al., 1998) and *Ece1* (Yanagisawa et al., 1998) clarified that homeotic transformation of the mandible into a maxilla had taken place. The role of *edn* in mandibular evolution was confirmed in zebrafish (Miller et al., 2000). The endothelin phenotypes are mediated by *dlx5* and *dlx6* transcription factors. Interestingly, $Dlx5^{-/-}$, $Dlx6^{-/-}$ double knockouts (Beverdam et al., 2002; Depew et al., 2002) have similar lower jaw morphology to the mice with knockouts of endothelin pathway molecules. In another set of experiments, the author showed that not only is endothelin necessary for mandibular patterning but ectopic expression of *Edn* in the domain of *Ednra* expression ($Ednra^{Edn1/+}$) changes the identity of the maxilla to mandible (Sato et al., 2008). Internally, the maxillary skeleton is not present and is replaced by a mandibular dentary bone and Meckel's cartilage. The maxilla does not form cartilage in normal development (Lee et al., 2004) so the presence of a cartilage rod is a major marker of change in identity. Endothelins are not expressed in other true organizers such as Hensen's node, the mid-hindbrain isthmus or polarizing region. Endothelin appears to be uniquely associated with craniofacial patterning.

7. Concluding remarks

In conclusion, we can add to the small list of bone fide craniofacial organizers by including the foregut endoderm (Benouaiche et al., 2008; Brito et al., 2006, 2008) and nasal pit (Szabo-Rogers et al., 2009) since both tissues direct the fate of the host facial mesenchyme (Table 1). Some of the key signals being secreted by craniofacial organizers include noggin, endothelin and Sonic hedgehog. Other signals may cooperate to complete patterning of the facial regions. In terms of broader capabilities, craniofacial organizer tissues have not yet been tested outside of the head as has been done for the anterior intestinal portal endoderm (Anderson et al., 2016). Organizers in general do share a common genetic signature enriched for signaling molecules and depleted of transcription factors (Anderson et al., 2016). Next on the agenda would be to profile transcriptome changes induced by a facial organizer then generate a gene regulatory network as was done for Hensen's node induction of neural plate (Trevers et al., 2023). Such experiments would lead to a deeper understanding of craniofacial patterning.

Acknowledgments

Work from the Richman lab was funded by the Canadian Institutes of Health Research grants to JMR and her trainees.

References

Abzhanov, A., & Tabin, C. J. (2004). Shh and Fgf8 act synergistically to drive cartilage outgrowth during cranial development. *Developmental Biology, 273*, 134–148.

Anderson, C., Khan, M. A. F., Wong, F., Solovieva, T., Oliveira, N. M. M., Baldock, R. A., ... Stern, C. D. (2016). A strategy to discover new organizers identifies a putative heart organizer. *Nature Communications, 7*, 12656.

Anderson, C., & Stern, C. D. (2016). Organizers in development. *Current Topics in Developmental Biology, 117*, 435–454.

Ashique, A. M., Fu, K., & Richman, J. M. (2002). Endogenous bone morphogenetic proteins regulate outgrowth and epithelial survival during avian lip fusion. *Development (Cambridge, England), 129*, 4647–4660.

Benouaiche, L., Gitton, Y., Vincent, C., Couly, G., & Levi, G. (2008). Sonic hedgehog signalling from foregut endoderm patterns the avian nasal capsule. *Development (Cambridge, England), 135*, 2221–2225.

Beverdam, A., Merlo, G. R., Paleari, L., Mantero, S., Genova, F., Barbieri, O., ... Levi, G. (2002). Jaw transformation with gain of symmetry after Dlx5/Dlx6 inactivation: Mirror of the past? *Genesis, 34*, 221–227.

Bhattacharyya, S., Bailey, A. P., Bronner-Fraser, M., & Streit, A. (2004). Segregation of lens and olfactory precursors from a common territory: Cell sorting and reciprocity of Dlx5 and Pax6 expression. *Developmental Biology, 271*, 403–414.

Bhattacharyya, S., & Bronner-Fraser, M. (2008). Competence, specification and commitment to an olfactory placode fate. *Development (Cambridge, England), 135*, 4165–4177.

Bothe, I., Tenin, G., Oseni, A., & Dietrich, S. (2011). Dynamic control of head mesoderm patterning. *Development (Cambridge, England), 138*, 2807–2821.

Brito, J. M., Teillet, M. A., & Le Douarin, N. M. (2006). An early role for sonic hedgehog from foregut endoderm in jaw development: Ensuring neural crest cell survival. *Proceedings of the National Academy of Sciences of the United States of America, 103*, 11607–11612.

Brito, J. M., Teillet, M. A., & Le Douarin, N. M. (2008). Induction of mirror-image supernumerary jaws in chicken mandibular mesenchyme by Sonic Hedgehog-producing cells. *Development (Cambridge, England), 135*, 2311–2319.

Candido-Ferreira, I. L., Lukoseviciute, M., & Sauka-Spengler, T. (2023). Multi-layered transcriptional control of cranial neural crest development. *Seminars in Cell & Developmental Biology, 138*, 1–14.

Celá, P., Buchtová, M., Veselá, I., Fu, K., Bogardi, J. P., Song, Y., ... Richman, J. M. (2016). BMP signaling regulates the fate of chondro-osteoprogenitor cells in facial mesenchyme in a stage-specific manner. *Developmental Dynamics, 245*, 947–962.

Clouthier, D. E., Hosoda, K., Richardson, J. A., Williams, S. C., Yanagisawa, H., Kuwaki, T., ... Yanagisawa, M. (1998). Cranial and cardiac neural crest defects in endothelin-A receptor-deficient mice. *Development (Cambridge, England), 125*, 813–824.

Clouthier, D. E., Williams, S. C., Yanagisawa, H., Wieduwilt, M., Richardson, J. A., & Yanagisawa, M. (2000). Signaling pathways crucial for craniofacial development revealed by endothelin-A receptor-deficient mice. *Developmental Biology, 217*, 10–24.

Couly, G., Creuzet, S., Bennaceur, S., Vincent, C., & Le Douarin, N. M. (2002). Interactions between Hox-negative cephalic neural crest cells and the foregut endoderm in patterning the facial skeleton in the vertebrate head. *Development (Cambridge, England), 129*, 1061–1073.

Couly, G., Grapin-Botton, A., Coltey, P., Ruhin, B., & Le Douarin, N. M. (1998). Determination of the identity of the derivatives of the cephalic neural crest: Incompatibility between Hox gene expression and lower jaw development. *Development (Cambridge, England), 125*, 3445–3459.

Couly, G. F., Coltey, P. M., & Douarin, N. M. L. (1993). The triple origin of skull in higher vertebrates: A study in quail-chick chimeras. *Development (Cambridge, England), 117*, 409–429.

Couly, G. F., Coltey, P. M., & Le Douarin, N. M. (1992). The developmental fate of the cephalic mesoderm in quail-chick chimeras. *Development (Cambridge, England), 114*, 1–15.

Depew, M. J., Lufkin, T., & Rubenstein, J. L. (2002). Specification of jaw subdivisions by Dlx genes. *Science (New York, N. Y.), 298*, 381–385.

Etchevers, H. C., Vincent, C., Douarin, Le, Couly, N. M., & F, G. (2001). The cephalic neural crest provides pericytes and smooth muscle cells to all blood vessels of the face and forebrain. *Development (Cambridge, England), 128*, 1059–1068.

Gendron-Maguire, M., Mallo, M., Zhang, M., & Gridley, T. (1993). Hoxa-2 mutant mice exhibit homeotic transformation of skeletal elements derived from cranial neural crest. *Cell, 75*, 1317–1331.

Gurdon, J. B. (1987). Embryonic induction-molecular prospects. *Development (Cambridge, England), 99*, 285–306.

Hu, D., Colnot, C., & Marcucio, R. S. (2008). Effect of bone morphogenetic protein signaling on development of the jaw skeleton. *Developmental Dynamics, 237*, 3727–3737.

Hu, D., & Helms, J. A. (1999). The role of sonic hedgehog in normal and abnormal craniofacial morphogenesis. *Development (Cambridge, England), 126*, 4873–4884.

Hu, D., & Marcucio, R. S. (2009). Unique organization of the frontonasal ectodermal zone in birds and mammals. *Developmental Biology, 325*, 200–210.

Hu, D., Marcucio, R. S., & Helms, J. A. (2003). A zone of frontonasal ectoderm regulates patterning and growth in the face. *Development (Cambridge, England), 130*, 1749–1758.

Koentges, G., & Lumsden, A. (1996). Rhombencephalic neural crest segmentation is preserved throughout craniofacial ontogeny. *Development (Cambridge, England), 122*, 3229–3242.

Kurihara, Y., Kurihara, H., Suzuki, H., Kodama, T., Maemura, K., Nagai, R., ... Kamada, N. (1994). Elevated blood pressure and craniofacial abnormalities in mice deficient in endothelin-1. *Nature, 368*, 703–710.

Le Lievre, C. S. (1978). Participation of neural crest-derived cells in the genesis of the skull in birds. *Journal of Embryology and Experimental Morphology, 47*, 17–37.

Lee, S. H., Bedard, O., Buchtova, M., Fu, K., & Richman, J. M. (2004). A new origin for the maxillary jaw. *Developmental Biology, 276*, 207–224.

Lee, S. H., Fu, K. K., Hui, J. N., & Richman, J. M. (2001). Noggin and retinoic acid transform the identity of avian facial prominences. *Nature, 414*, 909–912.

Miller, C. T., Schilling, T. F., Lee, K., Parker, J., & Kimmel, C. B. (2000). sucker encodes a zebrafish Endothelin-1 required for ventral pharyngeal arch development. *Development (Cambridge, England), 127*, 3815–3828.

Nimmagadda, S., Buchtova, M., Fu, K., Geetha-Loganathan, P., Hosseini-Farahabadi, S., Trachtenberg, A. J., ... Richman, J. M. (2015). Identification and functional analysis of novel facial patterning genes in the duplicated beak chicken embryo. *Developmental Biology, 407*, 275–288.

Noden, D. M. (1978). The control of avian cephalic neural crest cytodifferentiation. I. Skeletal and connective tissues. *Developmental Biology, 67*, 296–312.

Noden, D. M. (1983). The role of the neural crest in patterning of avian cranial skeletal, connective, and muscle tissues. *Developmental Biology, 96*, 144–165.

Noden, D. M., & Francis-West, P. (2006). The differentiation and morphogenesis of craniofacial muscles. *Developmental Dynamics, 235*, 1194–1218.

Ozeki, H., Kurihara, Y., Tonami, K., Watatani, S., & Kurihara, H. (2004). Endothelin-1 regulates the dorsoventral branchial arch patterning in mice. *Mechanisms of Development, 121*, 387–395.

Pizette, S., & Niswander, L. (2000). BMPs are required at two steps of limb chondrogenesis: Formation of prechondrogenic condensations and their differentiation into chondrocytes. *Developmental Biology, 219*, 237–249.

Richman, J. M., & Delgado, J. L. (1995). Locally released retinoic acid leads to facial clefts in the chick embryo but does not alter the expression of receptors for fibroblast growth factor. *Journal of Craniofacial Genetics and Developmental Biology, 15*, 190–204.

Rijli, F. M., Mark, M., Lakkaraju, S., Dierich, A., Dolle, P., & Chambon, P. (1993). A homeotic transformation is generated in the rostral branchial region of the head by disruption of Hoxa-2, which acts as a selector gene. *Cell, 75*, 1333–1349.

Ruest, L. B., Xiang, X., Lim, K. C., Levi, G., & Clouthier, D. E. (2004). Endothelin-A receptor-dependent and -independent signaling pathways in establishing mandibular identity. *Development (Cambridge, England), 131*, 4413–4423.

Sato, T., Kurihara, Y., Asai, R., Kawamura, Y., Tonami, K., Uchijima, Y., ... Kurihara, H. (2008). An endothelin-1 switch specifies maxillomandibular identity. *Proceedings of the National Academy of Sciences of the United States of America, 105*, 18806–18811.

Szabo-Rogers, H. L., Geetha-Loganathan, P., Nimmagadda, S., Fu, K. K., & Richman, J. M. (2008). FGF signals from the nasal pit are necessary for normal facial morphogenesis. *Developmental Biology, 318*, 289–302.

Szabo-Rogers, H. L., Geetha-Loganathan, P., Whiting, C. J., Nimmagadda, S., Fu, K., & Richman, J. M. (2009). Novel skeletogenic patterning roles for the olfactory pit. *Development (Cambridge, England), 136*, 219–229.

Trevers, K. E., Lu, H. C., Yang, Y., Thiery, A. P., Strobl, A. C., Anderson, C., ... Stern, C. D. (2023). A gene regulatory network for neural induction. *Elife, 12*.

Waddington, C. H. (1966). *Principles of development and differentiation*. New York: Macmillan.

Webb, J. F., & Noden, D. M. (1993). Ectodermal placodes—Contributions to the development of the vertebrate head. *American Zoologist, 33*, 434–447.

Yanagisawa, H., Yanagisawa, M., Kapur, R. P., Richardson, J. A., Williams, S. C., Clouthier, D. E., ... Hammer, R. E. (1998). Dual genetic pathways of endothelin-mediated intercellular signaling revealed by targeted disruption of endothelin converting enzyme-1 gene. *Development (Cambridge, England), 125*, 825–836.

CHAPTER FOUR

Organizing activities of axial mesoderm

Elizabeth Manning[a,b] and Marysia Placzek[a,b,c,*]

[a]School of Biosciences, University of Sheffield, Sheffield, United Kingdom
[b]Bateson Centre, University of Sheffield, Sheffield, United Kingdom
[c]Neuroscience Institute, University of Sheffield, Sheffield, United Kingdom
*Corresponding author. e-mail address: m.placzek@sheffield.ac.uk

Contents

1. The embryonic organizer, neural induction and secondary organizers	84
2. The role of axial mesoderm in the chick	87
2.1 Chick axial mesoderm development	87
2.2 Young head process mesoderm and its derivatives stabilize anterior neural fate	88
2.3 Chick prechordal mesoderm/mesendoderm acts as a local organizer along the A-P and D-V axes	91
2.4 Young head process mesoderm promotes morphogenesis	92
2.5 Chick notochord acts as a local organizer, potentially influencing all three axes	93
3. The role of axial mesoderm in Xenopus and zebrafish	95
3.1 Axial mesoderm development in Xenopus and zebrafish	95
3.2 In Xenopus and zebrafish, prechordal mesoderm/mesendoderm maintains A-P regional identity	96
3.3 Prechordal mesoderm/mesendoderm and notochord reveal organization through their impact on morphogenesis	101
3.4 In Xenopus and zebrafish, prechordal mesoderm/mesendoderm and notochord maintain regional identity along the D-V and medio-lateral axes	103
4. The role of axial mesoderm in mouse	104
4.1 Mouse axial mesoderm development	104
4.2 Mouse prechordal mesoderm/mesendoderm and notochord act as local organizers	104
4.3 Evolutionarily-conserved secreted factors mediate PMe and notochord activities	106
4.4 Prechordal mesoderm/mesendoderm-derived signaling factors	107
4.5 Notochord-derived signaling factors	109
5. Summary	112
References	112

Current Topics in Developmental Biology, Volume 157
ISSN 0070-2153, https://doi.org/10.1016/bs.ctdb.2024.02.007
Copyright © 2024 Elsevier Inc. All rights are reserved, including those for text and data mining, AI training, and similar technologies.

Abstract

For almost a century, developmental biologists have appreciated that the ability of the embryonic organizer to induce and pattern the body plan is intertwined with its differentiation into axial mesoderm. Despite this, we still have a relatively poor understanding of the contribution of axial mesoderm to induction and patterning of different body regions, and the manner in which axial mesoderm-derived information is interpreted in tissues of changing competence. Here, with a particular focus on the nervous system, we review the evidence that axial mesoderm notochord and prechordal mesoderm/mesendoderm act as organizers, discuss how their influence extends through the different axes of the developing organism, and describe how the ability of axial mesoderm to direct morphogenesis impacts on its role as a local organizer.

1. The embryonic organizer, neural induction and secondary organizers

In 1924, the pioneering studies of Hilde Mangold and Hans Spemann revealed the astonishing influence of the dorsal blastopore lip of amphibians on body plan development. Using two differently colored species of newts at early gastrula stages of development, they showed that transplanting cells from the morphologically discrete dorsal blastopore lip of one embryo to the ventral-marginal region of a host embryo resulted in the development of a secondary embryo, with coherent anterior-posterior (A-P) and dorso-ventral (D-V) axes, much of which—including most of the nervous system—derived from cells of the host (Spemann & Mangold, 1924). This cell population was named the embryonic organizer, or the Spemann organizer (Harland & Gerhart, 1997; and see accompanying article by Slack (2024)). Studies in rabbits and birds then revealed that the tip of the primitive streak is functionally equivalent to the Spemann organizer: when transplanted to a 'naïve' region of a host, the tip of the primitive streak is able to instruct host cells to change their fate and to direct development of the body axes (Anderson & Stern, 2016; Waddington & Schmidt, 1933; Waddington, 1933, 1934, 1937). These studies led to the concept of an 'organizer', an embryonic tissue with a dual function—the ability to change the fate of adjacent cells (termed 'induction'), and the ability to pattern adjacent cells, a process by which cells gain positional information and adopt specific fates in defined positions (Anderson & Stern, 2016; Gurdon, 1987). The ability of the embryonic organizer to induce a coherent nervous system with recognizable brain and spinal cord means that it is particularly associated with the ability to redirect the fate of non-neural ectoderm cells to a neural

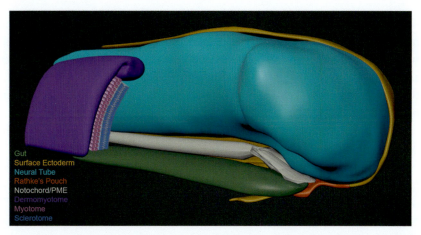

Fig. 1 Axial mesoderm occupies a central position within the developing body. Schematic showing the position of different body structures relative to the fan-shaped prechordal mesoderm/mesendoderm and rod-shaped notochord in the neurula stage embryo. In the head, the prechordal mesoderm/mesendoderm lies ventral to the forebrain, dorsal to the foregut, medial to cardiac precursors (not shown), and at its anterior limit, abuts placodal tissue that will give rise to Rathke's pouch, the precursor of the anterior pituitary. In neck regions, anterior notochord lies ventral to the midbrain, hindbrain and anterior-most spinal cord; in the trunk and tail, notochord lies ventral to the spinal cord, dorsal to organs such as the pancreas (not shown), and medial to mesoderm cells, including those that form the somites (precursors of muscles and cartilage) and kidneys (not shown).

identity (Harland & Gerhart, 1997; Hemmati-Brivanlou & Melton, 1997). This is called neural induction and classic views have imagined this as a single event, causing a switch of fate.

During gastrulation, cells at the dorsal blastopore lip/tip of the primitive streak (frequently termed the node/Hensen's node: see below and accompanying article by Stern (2024)) undergo remarkable changes in behavior and morphology. They involute, or ingress, and give rise to axial mesoderm, or axial mesendoderm (here, for simplicity, we refer to it as axial mesoderm), a midline rod of cells that extends along the A-P axis and occupies a central position within the developing body (Fig. 1). Axial mesoderm is made up of two fundamentally different tissues, namely prechordal tissues and chordamesoderm (the defining feature of chordates): the latter gives rise to notochord. Prechordal tissues and chordamesoderm emerge sequentially from the dorsal blastopore lip/primitive streak tip and form in a stereotypical anterior to posterior manner as gastrulation proceeds. By neurula stages, the mesenchymal-like prechordal tissues constitute the anterior-most axial mesoderm and the epithelial-like chordamesoderm/notochord constitutes the rest of the axial mesoderm. As axial mesoderm develops from embryonic organizer cells, the neural plate becomes overtly obvious and quickly assumes regional character along both its A-P and D-V axes, suggesting an intimacy and coherence in the development of axial mesoderm and the induction and patterning of neural tissue.

Grafting studies in newts and chicks showed that while the embryonic organizer from an early gastrula-stage embryo can direct formation of a complete axis, the embryonic organizer from a late gastrula-stage embryo (a stage at which prechordal tissues have emerged) orchestrates formation of just the trunk and tail (Dias & Schoenwolf, 1990; Nieuwkoop, 1997; Spemann, 1929; Spemann, 1931a, 1931b; Stern, 2005; Storey, Crossley, De Robertis, Norris, & Stern, 1992). Quickly, the idea emerged that the embryonic organizer might, in fact, harbor several separate organizers, each with distinct inducing and patterning properties. The idea was that the early dorsal blastopore lip contains all the organizers required to induce a complete axis, but that an 'anterior' or 'head' organizer is lost from the dorsal blastopore lip by late gastrulation. Supporting this idea, grafts of axial mesoderm from different A-P levels of a neurula-stage newt embryo were found to induce, specifically, the regions of the body to which they are normally adjacent (Mangold, 1933). Thus grafts containing prechordal tissues induced head and brain, while grafts containing chordamesoderm/notochord of different axial levels induced trunk and tail. This suggested that the head/brain are induced and patterned by prechordal tissues while more posterior parts of the axis/neuraxis are induced and patterned by chordamesoderm/notochord, each acting as a so-called 'secondary organizer' (Anderson & Stern, 2016; Mangold, 1933).

The overall central location of axial mesoderm within the wider forming body, including its intimate association with medial neural plate/ventral neural tube means that in theory it is superbly positioned to act as a secondary organizer of adjacent tissues, including neural tissue, and since Mangold's seminal work numerous studies in many different model organisms have investigated this possibility. As we discuss, the central dogma—the idea that prechordal tissues and notochord can induce and pattern different types of neural tissue along the A-P axis—has received very little support. Nonetheless, prechordal tissues and notochord can each act as local organizers, directing fate and pattern along the A-P and the D-V axes in specified neural tissue, and additionally directing fate and pattern in mesodermal and endodermal tissues. The extent to which they do so, however, is likely to vary in a species-dependent manner. Below, we focus on key lines of evidence gained in different species that examine whether prechordal tissues and notochord are local organizers. Finally, we briefly describe how molecular studies have both informed and challenged the concept of prechordal tissues and notochord as organizers.

2. The role of axial mesoderm in the chick

2.1 Chick axial mesoderm development

In the chick, as in all vertebrates, axial mesoderm is generated from anterior to posterior. Cells that ingress through the anterior tip of the extending primitive streak at Hamburger-Hamilton stage 2–3 (HH2–3) and the medial portion of Hensen's node (a bulbous thickening that forms at the tip of the fully extended primitive streak at HH4) migrate anteriorly, and by HH4+, form a broad short shaft of tissue that mingles with anterior endoderm and is termed young head process mesendoderm. At HH5, some 6 h after formation of Hensen's node, young head process mesendoderm begins to resolve into prechordal mesendoderm, prechordal mesoderm (which we refer to collectively in the text as PMe) that underlies the forebrain, and chordamesoderm that underlies the midbrain and hindbrain (Fig. 2). Cells that ingress through Hensen's node from HH5 give rise to

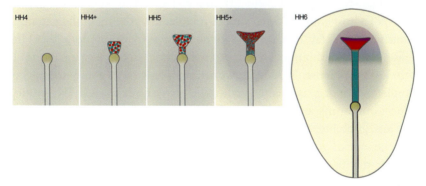

Fig. 2 Development of chick axial mesoderm. Left hand panels: Young head process mesendoderm, comprising a mix of anterior endoderm, PMe (red) and anterior chordamesoderm (turquoise) precursors begins to extend anteriorly from Hensen's node between HH4-HH4+. The PMe and anterior chordamesoderm begin to resolve over HH5-HH6: at this point the chordamesoderm begins to form the rod-like notochord. Right hand panel: Schematic of HH6 chick, showing position of PMe, notochord and Hensen's node relative to neural plate (gray) and epiblast (yellow). Regional identity is becoming apparent along the A-P axis of the neural plate.

notochord that extends from hindbrain to thoracic regions (Dias & Schoenwolf, 1990; Nicolet, 1971; Patten, Kulesa, Shen, Fraser, & Placzek, 2003; Schoenwolf & Sheard, 1990; Selleck & Stern, 1991; Storey et al., 1992). Finally, more posterior notochord develops from neuromesodermal progenitors (NMPs), bipotential stem-like cells that generate both neural and mesoderm tissue and that arise in the caudal lateral epiblast (a stem-like zone close to the remnants of Hensen's node) (Guillot, Djeffal, Michaut, Rabe, & Pourquié, 2021; Henrique, Abranches, Verrier, & Storey, 2015).

Non-ingressing epiblast cells give rise to neural tissue, and at least overtly, neural tissue is laid down at the same time that axial mesoderm forms, and, likewise, is laid down from anterior to posterior.

2.2 Young head process mesoderm and its derivatives stabilize anterior neural fate

There are significant challenges in determining whether a tissue has organizing activity and tests of organizer activity rely on a battery of approaches, key to which are manipulations of the tissue itself (Box 1). The chick provides an outstanding model to investigate whether axial mesoderm has organizing activity due to the ease with which PMe and notochord can be cleanly isolated and assayed, and the numerous conditional approaches that can be taken. In chick, organizer activity is classically assayed by grafting donor tissue to the area opaca (a region that lies outside the epiblast-forming area pellucida and will give rise to extra-embryonic tissues) of a host embryo. Such studies revealed that the tip of the extending primitive streak (at HH2–3) has organizing activity, as does the HH4 Hensen's node, although to a reduced extent (see accompanying chapter by Stern (2024)). Hensen's node, however, rapidly loses its ability to induce a complete secondary axis, and a graft of Hensen's node from a HH5 embryo induces only trunk and tail (Dias & Schoenwolf, 1990; Storey et al., 1992). The loss in ability of Hensen's node to induce head structures coincides with the exit of young head process mesendoderm raising the question of whether this tissue, or its descendants (PMe and anterior notochord) are 'head' organizers, while more posterior notochord is a 'trunk/tail' organizer. Studies focused especially on whether different regions of axial mesoderm are able to induce naive cells to a neural identity and impose neural regional pattern.

Potentially, the very earliest-generated axial mesoderm retains a very limited ability to induce neural tissue: young head process mesendoderm from HH4+ embryos can induce a neural plate-like structure when grafted

> **Box 1 Does a tissue act as an organizer?.**
> Significant challenges are faced in deciphering whether a tissue has organizing activity:
> 1. Organizing activity may be dispersed in time and space. Development is a dynamic process, in which the embryo undergoes significant changes in morphology, and tissues and cells change their relative positions and neighbors. It is possible, therefore, for a responding tissue to be exposed to an 'organizing' signal that derives from two separate tissues, each adjacent to the responding tissue at different times. For instance, many factors expressed by axial mesoderm are also expressed in the anterior endoderm. This means that to test if a defined tissue acts as an organizer in a classic grafting approach it must be accurately dissected, and not contaminated with an adjacent tissue that may have the same activity. At the same time, loss of function studies should be performed, to determine the requirement for the potential organizer; removal of a defined tissue, either physically or genetically, can be difficult to accomplish.
> 2. Organizing activity can only be revealed in tissues that are competent to respond, meaning that potential organizers must be grafted at a particular time and place. For instance, while transplantation of the mouse node to a mouse leads to an incomplete secondary axis that lacks the head and brain, transplantation of the mouse node to a chick leads to a complete secondary axis (see Martinez Arias & Steventon, 2018). The 'naïve' regions of the mouse and chick host embryos have a different ability to respond to, or interpret, the same instructions.
> 3. If an organizer unmasks a fate or pattern that is latent in the host tissue, it is said to 'evocate' a response in the host whereas if it an organizer redirects fate and pattern, it is said to 'induce' a response (discussed in Anderson & Stern, 2016; Martinez Arias & Steventon, 2018; see also accompanying article by Slack (2024)). As we discuss in the text, the question of whether an organizer 'induces' or 'evocates' a response is tied to the state of competence of a tissue, and the extent to which the tissue is already poised to adopt a particular fate.

to the area opaca of a host (Izpisúa-Belmonte, De Robertis, Storey, & Stern, 1993). Other studies, grafting HH5 young head process mesendoderm or HH5–6 PMe have argued that even as late as HH5–6, young head process mesendoderm, or PMe alone, can convert naive cells to an anterior neural identity (Pera & Kessel, 1997; Yoshihi et al., 2022). Grafts of HH5 young head process resulted in the appearance of ectopic brain-like structures, expressing markers of forebrain, midbrain and hindbrain (Yoshihi et al., 2022), while grafts of the PMe induced neural-like tissue expressing forebrain but not midbrain markers (Pera & Kessel, 1997). However, while an independent study similarly showed that HH5 quail PMe induced forebrain (but not midbrain) markers when grafted adjacent to the area pellucida, this study found no evidence that the PMe could divert area opaca cells to a neural fate (Foley, Storey, & Stern, 1997). A likely interpretation of these different outcomes is that the ability of grafts to induce neural tissue depends on the competence of the host tissue, and indeed, where HH5–6 young head process mesoderm or PMe is able to promote brain-like tissues, the grafts tend to be situated close to the boundary of the area opaca and area pellucida (Pera & Kessel, 1997), or even within the anterior epiblast itself and close to the developing host neural tube (Yoshihi et al., 2022). Studies that combine grafting experiments with transcriptomic analyses indicate that a pre-neural state is established in epiblast cells several hours before the overt appearance of neural tissue, or the formation of Hensen's node (Trevers et al., 2018; see accompanying chapter by Stern (2024)). This, together with older studies (Chapman, Brown, Lees, Schoenwolf, & Lumsden, 2004) suggests that signals deriving from the hypoblast (a layer of extraembryonic cells equivalent to the anterior visceral endoderm (AVE) of the mouse) or the definitive endoderm prime epiblast cells to a pre-neural state, where they are already poised to adopt anterior neural identity. As the hypoblast is displaced by definitive endoderm, it may continue to prime cells of the area pellucida/epiblast—but not the area opaca—to a pre-neural state. In this case, the HH5–6 young head process mesoderm, or PMe can evoke forebrain identity in tissue that is already poised to acquire neural identity, but is not capable of simultaneously initiating neural induction and patterning neural tissue, in other words, is not formally a 'brain organizer'. Recent studies, in which a chick Hensen's node is grafted to a naïve site, and the response of cells analyzed through single cell RNA-sequencing (scRNA-seq), suggest that neural induction is initiated rapidly, within an hour, but that a complex gene regulatory network then builds over the next 12 h before neural tissue is overtly obvious (Trevers et al., 2023). Other studies likewise suggest that several hours of exposure to 'organizer' signals is needed to stabilize neural identity in epiblast cells (Streit et al., 1998; Trevers et al., 2018; see also accompanying chapter by Stern (2024)). This prolonged requirement suggests therefore that young head process mesendoderm and PMe stabilize anterior neural identity, rather than switch epiblast cells to an anterior neural fate.

2.3 Chick prechordal mesoderm/mesendoderm acts as a local organizer along the A-P and D-V axes

Unquestionably, though, the PMe acts as a local organizer, and is able to switch the fate of (recently) specified neural cells, promoting both a more anterior and a more ventral identity. In the developing embryo, the PMe lies beneath the forming hypothalamus in the anterior-ventral neural tube (Burbridge, Stewart, & Placzek, 2016; Chinnaiya et al., 2023; Dale et al., 1997, 1999; Kim et al., 2022; Seifert, Jacob, & Jacob, 1993), and a wealth of evidence points to its key role in establishing hypothalamic fate. When grafted next to prospective hindbrain tissue, or grafted to the area opaca together with a HH5 node, the PMe upregulates *OTX2*, a gene that is widely expressed in the forebrain, including the hypothalamus (Foley et al., 1997), and when grafted to the area pellucida, HH6 PMe stabilizes neural tissue that specifically expresses the hypothalamic marker, *NKX2-1* (Pera & Kessel, 1997). Tissue recombinates, in which the PMe is apposed to neural explants of prospective hindbrain, likewise show that the PMe upregulates the hypothalamic-characteristic genes, *SHH* and *NKX2-1* (Dale et al., 1997, 1999; Ohyama, Ellis, Kimura, & Placzek, 2005). The PMe can therefore switch the fate of hindbrain neural tissue to a hypothalamic identity, and promote hypothalamic identity in nascent neural tissue. The competence of prospective hindbrain to respond to the PMe declines by HH7–9 (Foley et al., 1997), but beyond this stage, the PMe may continue to be able to divert the fate of other brain regions. RNA velocity plots from scRNA-seq datasets, a computational method that suggests lineage trajectories, reveals a conversion of prethalamic tissue (a relatively posterior-dorsal tissue) to hypothalamic tissue in early neurula stages of development, a time when the PMe still lies beneath the forming hypothalamus (Kim et al., 2022).

At the same time, the PMe imparts fine-grained positional information to specify hypothalamic cells along the D-V axis. Tissue recombinates, in which PMe is apposed to naïve neural explants, show that *SHH*, *NKX2-1* and *BMP7*—ventral-most markers that characterize hypothalamic floor plate-like cells (previously termed rostral diencephalic ventral midline cells) (Dale et al., 1997) are induced immediately adjacent to the PMe, but that the ventro-lateral markers, *LHX1* and *NKX2-2* are induced beyond these (Ohyama et al., 2005). Similarly, grafts of HH5 PMe to the area pellucida lead to neural tissue that expresses both the ventral hypothalamic marker, *NKX2-1* and the more dorsal forebrain marker *cNOT1* (Pera & Kessel, 1997). Intriguingly, though, HH6 PMe promotes neural tissue that expresses *NKX2-1* exclusively, suggesting that the D-V patterning ability of the PMe is extremely transient. Nonetheless, together these studies show that the PMe imparts anterior-ventral forebrain identity on more posterior-dorsal tissues.

Loss of function studies show the significance of the PMe in imparting anterior-ventral hypothalamic identity. Removal of the young head process mesendoderm at HH4+, or of the PMe (or PMe and anterior chordamesoderm) at HH5 does not lead to a complete loss of anterior neural tissue: the pan-anterior or dorsal markers, *OTX2, TAILLESS* and *PAX6* are still detected. However, the neural tube is shorter along its A-P axis and there is a loss or significant reduction in anterior ventral neural tissue, manifest through the reduced width of the optic vesicles (sometime resulting in cyclopia) and the reduction or loss of the hypothalamic markers, *SHH* and *NKX2.1* (Patten et al. 2003; Pera & Kessel, 1997).

2.4 Young head process mesoderm promotes morphogenesis

Fate-mapping experiments point to the importance of a prolonged ability of the PMe to impose regional anterior-ventral character. Both axial mesoderm and neural tissue undergo profound changes in morphogenesis over gastrula-to-neurula stages, each lengthening and thinning. Importantly, as this occurs, the PMe is not in simple register with neural cells, which instead move over and anterior to the PMe from more posterior regions in a conveyor-belt like manner (Burbridge et al., 2016; Chinnaiya et al., 2023; Dale et al., 1997; Rowan, Stern, & Storey, 1999). Thus, neural cells that were posterior-dorsal to the PMe at HH5–7 are briefly in register with it at HH8–9 and then migrate anterior to it at HH10, at which point the PMe is now in register with even more posterior-dorsal neural cells. Recent analyses, combining grafting studies with real-time analyses of cell movements suggest that the anterior movement of neural cells may be driven by young head process mesendoderm itself: when placed in the anterior epiblast, grafts containing these cells promote the convergence and migration of neural cells and epiblast cells towards them, resulting in an organized secondary anterior neuraxis, conjoined to that of the host (Yoshihi et al., 2022). (Note, this ability to promote migration again questions whether HH5–6 young head process mesoderm 'induces' ectopic brain structures: it is formally possible that the ectopic brain cells derive from host neural tissue itself).

In summary, the chick PMe acts as a local organizer of neural tissue along both the A-P and the D-V axes, able to switch fate and provide positional information. Whether it acts more widely to organize other local tissues, in other germ layers, remains unclear. In the embryo the PMe abuts placodal tissue that will give rise to Rathke's pouch (and hence the anterior pituitary), and lies medial to head and cardiac mesenchyme. Studies hint that the PMe may play a role in the development of these tissues (Gleiberman, Fedtsova, & Rosenfeld, 1999; Serrallach, Rauch, Lyons, & Huisman, 2022), but future studies are needed to confirm these possibilities. The signals that are likely to contribute to the ability of the PMe to organize adjacent tissues in an integrated manner along the A-P and D-V axes are discussed at the end of this review.

2.5 Chick notochord acts as a local organizer, potentially influencing all three axes

The different origins of anterior and posterior notochord (see above) raise the question of whether, like the PMe, they are able to direct early-specified neural tissue to a particular regional identity, or re-direct neural fates along the A-P axis. A handful of studies suggest that anterior-most notochord influences midbrain identity: when grafted to the area pellucida, HH5–6 anterior notochord stabilizes neural-like tissue that expresses the midbrain marker, Engrailed, but not forebrain markers (Pera & Kessel, 1997). Ex vivo studies show, further, that anterior notochord is able to confer relatively posterior-ventral identity on prospective forebrain tissue, upregulating the posterior ventral marker, Nkx6.1 (Qiu, Shimamura, Sussel, Chen, & Rubenstein, 1998), and repressing forebrain identity (Rowan et al., 1999), although the ability of notochord to influence A-P pattern is highly stage-dependent (in terms both of notochord and host). However, other studies, including those that extirpate the anterior notochord do not support the idea that anterior notochord is essential for the establishment of A-P pattern in the neural plate (Darnell, Schoenwolf, & Ordahl, 1992; Rowan et al., 1999), and currently, there is no evidence the posterior notochord, deriving from NMPS, imparts A-P regional identity on adjacent tissues. In summary, in contrast to the PMe, the notochord does not appear to be essential for A-P regional identity.

Certainly, though, throughout its entire length, the notochord induces cell fates and promotes patterning along the D-V axis, and can do so in tissues that derive from two germ layers—neural ectoderm and mesoderm. The ability of the notochord to influence fate and pattern along the D-V axis has been extensively reviewed (Anderson & Stern, 2016; Corallo, Trapani, & Bonaldo, 2015; Placzek & Briscoe, 2018), and here we provide just an overview.

Throughout the neural tube, distinct classes of neuronal progenitors are specified at different D-V positions. Progenitors are distributed bilaterally with an organization that is symmetric with reference to floor plate cells at the ventral midline and roof plate cells at the dorsal midline. At spinal cord levels, progenitors give rise to interneurons and motor neurons that mediate sensory-motor behaviors. A series of pioneering manipulations in the developing spinal cord demonstrated that the notochord controls the pattern of cell differentiation along the D-V axis. Transplantation of an ectopic notochord adjacent to the neural tube of a host embryo (thoracic levels for both graft and host) changes the pattern of cell differentiation of adjacent host cells. Floor plate and ventral motor neurons are induced in regions of the neural tube that are normally fated to give rise to dorsal roof plate cells and dorsal interneurons. Induced cells are arrayed in a manner that recapitulates their pattern in the normal ventral neural tube: ectopic floor plate cells differentiate immediately adjacent to grafted cells, whereas motor neurons are located at a characteristic distance (Placzek, Tessier-Lavigne, Yamada, Jessell, & Dodd, 1990; Placzek, Yamada, Tessier-Lavigne, Jessell, & Dodd, 1991; van Straaten, Hekking, Wiertz-Hoessels, Thors, & Drukker, 1988; Yamada, Placzek, Tanaka, Dodd, & Jessell, 1991). Conversely, notochord removal resulted in the absence of the floor plate and motor neurons (van Straaten et al., 1988; Yamada et al., 1991). Similarly the notochord can induce and pattern the D-V axis at hindbrain levels. The type of neuron that is induced reflects an A-P tissue competence: thus, for instance, serotonergic neurons are induced in the hindbrain (Yamada et al., 1991).

In conclusion, the chick notochord is able to both divert cell fate and to pattern cells in the prospective spinal cord and hindbrain and therefore functions as a bona fide organizer. This ability, nonetheless, is constrained through the competence of the host tissue. For instance, the notochord is able to induce neuronal fates only when grafted to neural tissue—and when grafted to other structures (such as the limb bud) instead evocates alternate fates (Johnson, Riddle, Laufer, & Tabin, 1994). Even within neural tissue, there is a temporal aspect to the ability of the notochord to induce and pattern: a notochord transplanted adjacent to the neural tube of older embryos induces motor neurons without prior induction of a floor plate (Yamada et al., 1991. Thus, the ability of the notochord to act as a local organizer is highly dependent on the competence of the neural tissue.

Likewise, although the notochord is fundamental to the D-V organization of the ventral neural tube, additional tissues contribute. Most obviously, floor plate cells that are induced by the notochord have similar local D-V organizing activity to notochord after transplantation (Placzek et al., 1991; Yamada et al., 1991). And while floor plate cells can be induced through vertical signals provided by the notochord, they can eventually differentiate after notochord ablation, and most likely do so under the direction of signals that derive from non-affected adjacent floor plate cells that operate in the plane of the neural tube (Artinger & Bronner-Fraser, 1993).

In the same vein, the notochord appears to act as an organizer of mesodermal tissues, in particular of the somites that lie lateral to the posterior neural tube. Each somite is composed of ventral sclerotome (future skeletal elements of the ribs and vertebral column) and dorsal dermomyotome (future dermis and myotome). The myotome in turn gives rise to medial epaxial muscle and lateral hypaxial muscle. Transplantation of an ectopic notochord adjacent to the lateral region of a host somite induces sclerotome fate, and in the mytotome, specifies medial epaxial muscle as opposed to lateral hypaxial muscle derivatives (Brand-Saberi, Ebensperger, Wilting, Balling, & Christ, 1993; Marcelle, Stark, & Bronner-Fraser, 1997; Pourquié, Coltey, Teillet, Ordahl, & Le Douarin, 1993; Pownall, Strunk, & Emerson, 1996). In the mesoderm, therefore, notochord influences fate along both the D-V and medio-lateral axes.

The notochord exerts a more widespread influence, affecting the development of additional adjacent tissues, but in many of these cases it does not act as an organizer. For instance, the notochord influences development of endodermal pancreas. The pancreas forms from the fusion of distinct dorsal and ventral buds, and the notochord is required for appropriate development of the dorsal bud. Removal of the notochord leads to inappropriate appropriate morphogenesis and differentiation of the dorsal bud (Kim, Hebrok, & Melton, 1997). However, in vitro recombination experiments suggest that the notochord can initiate and maintain dorsal bud identity only in pre-pancreatic endoderm, and not in non-pancreatic endoderm, and so evokes dorsal bud identity in tissue that has already been induced to a pre-pancreatic fate (Kim et al., 1997).

The signaling milieu provided by the notochord is complex, but there is widespread agreement that Shh mediates its ability to direct D-V pattern, and its mechanism of action is discussed at the end of this review.

3. The role of axial mesoderm in Xenopus and zebrafish

3.1 Axial mesoderm development in Xenopus and zebrafish

In Xenopus, as in newts, cells at the dorsal marginal zone of the early gastrula act as an 'embryonic organizer' and induce a complete axis when grafted ectopically to the ventral marginal zone (Martinez Arias & Steventon, 2018; Sander & Faessler, 2001; Shih & Keller, 1992). Axial mesoderm develops from cells with organizer activity and forms from anterior to posterior in a manner that reflects the time at which cells involute at the dorsal blastopore lip. Deep/ventral portions of the dorsal marginal zone involute first, and give rise to prechordal tissues, composed anteriorly of 'leading edge mesendoderm' (a tissue that expresses *Xhesx1*, potentially more similar to anterior endoderm), then *Xgsc*-expressing PMe. Superficial portions of the dorsal marginal zone then involute and give rise to *Xtbxt* (previously termed *Xbra*)-expressing chordamesoderm (future notochord) (Fig. 3) (Cho, Blumberg, Steinbeisser, & de Robertis, 1991; Dale & Slack, 1987; Keller, 1975, 1976).

In zebrafish, axial mesoderm precursor cells are more widely distributed in the late blastula-early gastrula embryo. The PMe arises from deep layers of the morphologically-distinctive shield at the dorsal margin, and anterior notochord from superficial layers. However, posterior notochord arises from more ventral regions of the margin (Fauny, Thisse, & Thisse, 2009; Kimmel, Warga, & Schilling, 1990; Thisse & Thisse, 2015). Thus, each part of the zebrafish margin forms axial mesoderm of a particular A-P identity. As in Xenopus, cells at the dorsal margin that give rise to prechordal tissues are the first population to ingress and collectively migrate toward the animal pole of the gastrula (future head region), followed by cells that form the notochord (Schulte-Merker et al., 1994).

3.2 In Xenopus and zebrafish, prechordal mesoderm/mesendoderm maintains A-P regional identity

In Xenopus and zebrafish, molecular and genetic experiments suggested that the PMe may play a role in head development. A cocktail of secreted factors, including BMP and Wnt antagonists were found to be expressed in the organizer/shield, and then in its axial mesoderm derivatives (De Robertis, Blum, Niehrs, & Steinbeisser, 1992; Jones & Mullins, 2022; Leyns, Bouwmeester, Kim, Piccolo, & De Robertis, 1997; Piccolo et al., 1999), and the combined activity of BMP and Wnt antagonism promoted the development of anterior neural fate in ectoderm (Hemmati-Brivanlou & Melton, 1997; Levine & Brivanlou, 2007; Wilson & Houart, 2004). A seminal finding was that the secreted Wnt antagonist, *dickkopf-1 (Xddk1/dkk1)* is expressed in the organizer/shield, and then in the PMe, and that overexpression of *Xdkk1* in Xenopus, or of *dkk1b* in zebrafish, promotes anterior

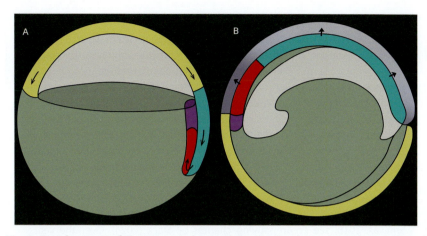

Fig. 3 Development of Xenopus axial mesoderm. Prechordal mesendoderm (red) and chordamesoderm (turquoise) are generated from cells that pass sequentially through the embryonic organizer. The first cells to traffic through the organizer of the early gastrula embryo will form the prechordal mesendoderm (PMe): at its anterior limit, the PMe abuts and merges into anterior endoderm tissues (*purple*). Chordamesoderm is then generated from cells that traffic through the organizer. Arrows in A show relative tissue movements. Arrows in B show vertical signaling from axial mesoderm to overlying neural plate (*gray*).

neural cell fates instead of posterior neural cell fates (Glinka et al., 1998; Hashimoto et al., 2000; Shinya et al., 2000; Tanaka, Hosokawa, Weinberg, & Maegawa, 2017), whereas *dkk1b* morphants are headless (Caneparo et al., 2007; Shinya et al., 2000). Such studies provided a rationale for why the PMe may be critical to head/brain development (Kiecker & Niehrs, 2001). At the same time, zebrafish *flh*, *mom*, *ntl* (the zebrafish homologue of *tbxt*) and *doc* mutants lack notochord and develop without trunk/tail structures, indicating that the notochord may play a role in trunk/tail development (Odenthal et al., 1996). Additionally, numerous genes—many of which are Nodal target genes (see Box 2)—were found to be expressed in sub-populations of organizer cells and then in particular domains of axial mesoderm, and when over-expressed or knocked-out, to affect A-P neuraxis development. Seminal work in Xenopus showed that *Xgsc* (which is expressed in the organizer and then in the PMe) has the remarkable ability to mimic Spemann's organizer in gain-of-function experiments: microinjection of *Xgsc* mRNA into the ventral region of a Xenopus blastula results in formation of a complete secondary axis (Cho et al., 1991). Subsequent work in zebrafish showed similarly, that *gsc* can induce a secondary axis in zebrafish, and recent work in

Box 2 A Nodal gradient directs axial mesoderm identity

Studies in Xenopus, zebrafish and mouse have provided an understanding of how axial mesoderm develops into PMe versus chordamesoderm/notochord. Seminal work in Xenopus suggested that PMe and chordamesoderm notochord development is initiated in the late blastula/early gastrula embryo through the graded activity of members of the TGF-beta superfamily factors, including Activin, Nodal and Nodal-like proteins: high concentrations of Activin/Nodal/Nodal-like proteins upregulate short-range target transcription factors that include *Xgsc*, while genes that include *Xtbxt* are induced by lower levels of Activin/Nodal/Nodal-like Nodal and increase in expression over time (reviewed in Barratt, Drover, Thomas, & Arkell, 2022; Dubrulle et al., 2015; Kiecker & Niehrs, 2001; Robertson, 2014; Schier & Shen, 2000; Stower & Srinivas, 2018; see accompanying chapter by Slack (2024). Over the years, transcriptomic studies have identified numerous additional genes whose expression overlaps with *Xgsc* and *Xtbxt*, all of which suggest that, although confined to the dorsal marginal zone, the Xenopus early gastrula organizer is already subdivided into PMe and chordamesoderm precursors. In zebrafish, fate appears to be determined through the ratio of Nodal and BMP signaling: a high Nodal: BMP ratio directs PMe identity, and a low Nodal: BMP ratio directs notochord identity. Thus, *gsc* is expressed in the presence of high pSmad2 (a Nodal signaling effector) and low/no pSmad5 (a BMP signaling effector) (Soh, Pomreinke, & Müller, 2020). As a consequence, *gsc* is expressed in deep/ventral parts of the embryonic organizer/shield, while *ntl (tbxt)* is expressed more widely. Transplantation studies showing that the deep *gsc*-expressing cells within the shield region are fated to give rise to PMe, whereas *ntl (tbxt)*-expressing superficial cells are fated to give rise to notochord, are can consistent with the idea that the early patterning of the mesoderm by Nodal begins to generate axial mesoderm pattern. Recent single cell RNA-seq (scRNA-seq) studies, analysing the response of zebrafish animal pole explants, confirm and extend these findings, showing that a regional Nodal gradient can generate a structure that recapitulates the key cell movements of gastrulation and differentiates into PMe, notochord, and tailbud-like cells along an A-P axis (Cheng et al., 2023). The duration of Nodal signaling, likewise, affects cell fate specification, and extended Nodal signaling within the shield promotes PMe specification (*gsc* expression) which acts as a transcriptional repressor, restraining Nodal signaling from upregulating the endoderm differentiation gene *sox17* within these cells (Sako et al., 2016). In summary, graded activity of Nodal directs axial mesoderm identity, higher levels of Noda/longer duration promoting PMe identity and lower levels/shorter duration promoting notochord.

Genetic studies in mouse revealed that, as in other vertebrates (see above) graded Nodal activity governs PMe and notochord identities, in mouse,

> acting in combination with the TGF-β related signaling molecule, GDF1 (growth and differentiation factor 1) (see Barratt et al., 2022). A peak of Nodal activity initiates formation of the anterior primitive streak; as gastrulation proceeds, Nodal activity diminishes in the anterior primitive streak, to a point where, acting with GDF1, it induces the node. A complete loss of Nodal activity prevents primitive streak formation and consequently, no axial mesoderm forms; a mild decrease in Nodal activity allows the node and the notochord, but not the PMe to form. These results are consistent with the idea that the PMe forms in response to higher levels of Nodal activity than are required for notochord formation, and forms prior to development of the morphologically obvious node.

zebrafish has identified a host of Nodal-target genes, including the BMP antagonist, *chordin (chrd)*, as well as *foxa3, lhx1a, tbr1b, pitx2, foxa* and *elovl6*, each able to induce different portions of the axis when overexpressed (Cheng et al. 2023); conversely, *ripply1*, an adaptor protein that recruits the global corepressor Groucho/TLE to T-box proteins and so antagonizes the transcriptional activation of *ntl (tbxt)*, induces embryos that lack a notochord and posterior axis (Cheng et al. 2023): classic studies had shown that mutations in zebrafish *ntl (tbxt)* lead to the failure of notochord development and loss of posterior structures (Odenthal et al., 1996).

There is no evidence to suggest that, in the gastrula embryo, involuted PMe or notochord act as classic 'head' or 'trunk/tail' organizer (i.e. able to induce and pattern cells to a regional neural identity). Instead, all the evidence points to head and trunk organization occurring prior to gastrulation, and mediated by a range of tissues, that, nonetheless, include PMe and notochord precursors. In Xenopus, organizer activity is typically assayed by grafting donor tissue to the ventral marginal zone (note other variations are to implant tissue into the blastocoel of a host ('Einsteck grafts'), or to set up ex vivo in recombinates with early gastrula animal cap ectoderm), and in zebrafish, organizer activity is assayed by grafting tissue into the animal pole. Classic studies in Xenopus and zebrafish, dissecting and analysing sub-regions of the organizer, showed that deep regions of the organizer/shield (fated to give rise to PMe) predominantly induce head structures whereas superficial layers (fated to give rise to notochord) mainly induce headless trunks (Saúde, Woolley, Martin, Driever, & Stemple, 2000; Zoltewicz & Gerhart, 1997).

Ablation experiments likewise indicate that in Xenopus, neural induction and A-P regionalization are dependent on PMe and notochord precursors (and potentially, other tissues that are mixed with them): extirpation of PMe or notochord precursors results in the loss of anterior and posterior tissues, respectively (Schneider & Mercola, 1999). Importantly, though, the information required to induce A-P regionalized neural tissue is passed swiftly to ectoderm, prior to the involution of axial mesoderm. Cultured explants of the Xenopus dorsal marginal zone 'Keller' explants) serve as a central paradigm to decipher whether neural tissue is induced and patterned through planar or vertical signals (the explants develop in such a manner that axial mesoderm and neural tissue are not vertically aligned, but instead aligned in the same plane) (Keller & Danilchik, 1988). Transcriptomic analyses of cultured Keller explants suggest that neural tissue, with the full extent of A-P pattern can develop in the absence of vertical signals (Kakebeen et al., 2021). This supports earlier work analysing explants or 'exogastrulae' (another method of Xenopus culture that maintains only planar contact), which showed that a significant amount of neural tissue, regionalized along the A-P axis, can be detected in the absence of vertical signaling (Doniach, Phillips, & Gerhart, 1992; Ruiz i Altaba, 1990). Likewise, in zebrafish, ex vivo assays suggest that at least some element of forebrain identity can occur in the absence of the PMe (Grinblat, Gamse, Patel, & Sive, 1998).

Genetic gain-of-function and loss-of-function experiments likewise show that in zebrafish, neural tissue can be induced and broadly regionalized along the A-P axis in a manner that is independent of axial mesoderm. For instance, zebrafish double mutants for the nodal-related genes *squint* and *cyclops* (*sqt;cyc*), as well as mutants for the essential Nodal cofactor *one-eyed pinhead* (*oep*), do not form a PMe but have a surprisingly normal A-P neural pattern (Gritsman, Talbot, & Schier, 2000; Schier, Neuhauss, Helde, Talbot, & Driever, 1997)—albeit that, similar to exogastruale, they are cyclopic, pointing to the importance of the PMe in ventral forebrain morphogenesis. Likewise, while zebrafish embryos mutant for the Wnt antagonist, *dkk1*, are headless, this likely reflects actions of dkk1 on both axial mesoderm precursors and neural tissue (Hashimoto et al., 2000). Further, experiments in zebrafish that accurately manipulate the timing of BMP inhibition show that different types of anterior neural tissue are specified sequentially through BMP inhibition prior to gastrulation, and a current idea is that neural induction and regionalization occur as ectoderm progressively loses competence to be able to adopt successive anterior (forebrain, then midbrain, then hindbrain) identities (Tucker, Mintzer, & Mullins, 2008). In this view neural tissue is progressively patterned along its

A-P axis over time through the action of BMP and Wnt antagonists that trigger a response as a function of the competence of host cells (see Andoniadou & Martinez-Barbera, 2013; Fekany-Lee, Gonzalez, Miller-Bertoglio, & Solnica-Krezel, 2000; Levine & Brivanlou, 2007). PMe and notochord precursors can provide such antagonists, but other tissues can contribute.

Nonetheless, in both Xenopus and zebrafish, once axial mesoderm has involuted, the PMe and notochord maintain expression of numerous factors implicated in anterior or posterior specification. The PMe, for instance, maintains expression of the transcription factors, *Xgsc/gsc, Xlim/lim1*, as well as secreted Wnt inhibitors such as *frzb1, dkk1* and *frzb2/crescent* (e.g. Niehrs, 1999; Tanaka et al., 2017). As in the chick, such factors maintain regional identity in induced neural tissue along the A-P axis. In zebrafish, the Wnt antagonist, *dkk1* is required at mid-to late gastrula stages (ie after axial mesoderm has involuted) for the maintenance of normal patterned heads, and neural cells lose competence to respond to *dkk1* only in early neurula stages of development (Kim et al., 2009; Shinya et al., 2000).

Likely, then, transcription factors that can induce a complete or partial neuraxis begin to operate prior to gastrulation, and then continue to exert an influence through their activities in axial mesoderm post-gastrulation. Transcriptomic studies indicate a sophisticated gene regulatory network (GRN) within the Xenopus and zebrafish organizer, into which such transcription factors genes, can feed (Blitz & Cho, 2021; Charney, Paraiso, Blitz, & Cho, 2017; Farrell et al., 2018; Kakebeen et al., 2021; Kiecker, Bates, & Bell, 2016; Popov et al., 2017). For instance, the role of *Xgsc/gsc* in promoting an axis is in part explained by its well characterized ability to transcriptionally activate or repress target genes within this network, including Wnt8 and BMP pathway components and their antagonists (Charney et al., 2017; Kumar, Umair, Lee, & Kim, 2023; Umair et al., 2021). Unraveling the details of this GRN will be essential in understanding how the PMe and notochord maintain A-P regional identity.

3.3 Prechordal mesoderm/mesendoderm and notochord reveal organization through their impact on morphogenesis

Studies in Xenopus and zebrafish show how the PMe and notochord reveal organization along the A-P axis through their impact on morphogenesis. Amongst the transcription factors that are maintained in the PMe are those that play a dual role, simultaneously affecting cell fate and morphogenesis. As mentioned above, the initial transcription of *Xgsc* in the organizer tissue itself is very transient, and by neurula stages, *Xgsc* persists in the PMe but ceases in resident organizer tissue and notochord (De Robertis, Fainsod, Gont, & Steinbeisser, 1994). PMe and chordamesoderm are not yet fully-committed as they first emerge from the organizer and resolve through self-organized reciprocal molecular mechanisms. In Xenopus, Gsc promotes PMe identity and simultaneously represses expression of *Xtbxt*, hence suppressing notochord identity (Artinger, Blitz, Inoue, Tran, & Cho, 1997). At the same time, Xtbxt, working indirectly, represses *Xgsc*. As Xenopus organizer-derived cells gradually segregate into these two cell populations, their characteristic cell behaviors become increasingly obvious. *Xgsc* promotes anterior cell migration, while *Xtbxt* promotes the convergent extension movements (medio-lateral cell intercalation) that result in formation of the notochord and lengthening of the A-P body axis (Domingo & Keller, 1995; Kwan, 2003; Winklbauer, 1990; Yamada, 1994). Other PMe-derived factors contribute to this: *Xcrs/Xfrzb2* not only blocks posteriorizing Wnts such as XWnt8 but also interferes with signals of the Wnt11 class that are required for gastrulation movements (Niehrs, 2022; Sokol, 2000). In zebrafish, scRNA-seq studies similarly show a subset of cells that co-express both PMe and notochord markers, arguing that axial mesoderm cells retain a plasticity as they first form (Farrell et al., 2018. Studies in zebrafish likewise show that *gsc* inhibits Wnt/PCP-mediated convergent extension (Ulmer et al., 2017), so convergent extension movements become more pronounced as gastrulation proceeds and the PMe and notochord resolve. In summary, Wnt inhibitors deriving from the PMe may simultaneously affect A-P specification and cellular movements. The impact of these factors on both signaling and morphogenesis suggests how earlier 'organization' events are revealed as axial mesoderm differentiates into PMe and notochord. Different A-P regions of the neuraxis begin to be induced and patterned within the organizer, then, as axial mesoderm emerges, it progressively reinforces pattern along the A-P axis, and determines the timing of the convergence and extension movements that elongate the body axis, and make A-P patterning globally more obvious.

At the same time, studies in zebrafish indicated that the resolution of PMe and notochord may impact on tissues more widely along the A-P axis. Cardiac progenitor cells reside in the region of lateral plate mesoderm that lies adjacent to the PMe-notochord junction. Laser ablation of the notochord tip causes a posterior expansion of *nkx2.5*, the earliest marker of presumptive myocardial cell fate (Goldstein & Fishman, 1998). Similarly, *nkx2.5* expression expands posteriorly in the *ntl* mutant. Therefore, the separation of PMe and notochord defines the extent of the heart field.

3.4 In Xenopus and zebrafish, prechordal mesoderm/mesendoderm and notochord maintain regional identity along the D-V and medio-lateral axes

In both Xenopus and zebrafish, the PMe and the notochord are implicated in the D-V patterning of the neural tube. Surgical or genetic ablation of the PMe results in D-V patterning defects in the developing forebrain and cyclopia—a failure of eye morphogenesis or development (Li, Tierney, Wen, Wu, & Rao, 1997; Ruiz i Altaba, 1990), while ablation of the notochord impacts on D-V patterning in spinal cord regions (Moghadam, Chen, & Heathcote, 2003). Such studies suggest that only brief periods of exposure to axial mesoderm are needed to initiate D-V pattern. Other studies suggest, by contrast, that the role of axial mesoderm is to maintain, not induce D-V pattern: in Xenopus, cultured Keller explants initially show signs of D-V pattern but this declines over time (Kakebeen et al., 2021), suggesting that signaling from underlying tissues is critical for the maintenance, but not initial onset of D-V pattern. Potentially, the requirement for axial mesoderm in D-V patterning is different at different A-P axial levels.

The notochord, additionally, promotes regional identity within somites. In zebrafish, there is virtually no sclerotome, but the notochord induces slow muscle fate from myotome cells, defining an 'adaxial' population (Blagden, Currie, Ingham, & Hughes, 1997). Many zebrafish mutants affecting notochord development, such as *flh*, *mom*, *ntl* and *doc*, display a loss of muscle pioneers, and cell transplantation experiments carried out in *ntl* and *doc* mutants showed that somite defects can be rescued by exogenous wild-type notochord cells (Halpern, Ho, Walker, & Kimmel, 1993; Odenthal et al., 1996; Talbot et al., 1995). Whether the notochord acts as an organizer of somite differentiation, however, is debatable: the correct specification of different cell types within the myotome is also dependent on the timing at which they receive the signal from the notochord and on their competence to respond (Wolff, Roy, & Ingham, 2003).

Embryological manipulations in amphibians have demonstrated that the notochord also plays a critical role in the development of the hypochord, a transient, endoderm-derived structure that lies immediately ventral to the notochord in the amphibian and fish embryo. Removal of the notochord during early neurulation leads to a complete failure of hypochord development and conversely, grafting experiments demonstrate that higher levels of notochord-derived signaling can increase the size of the hypochord (Cleaver, Seufert, & Krieg, 2000). Likewise, in *flh* and *ntl* zebrafish mutants which lack a notochord, the hypochord does not develop or it is severely disrupted (Halpern et al., 1995). Finally, work in zebrafish indicates that, as in chick, the notochord is required for development of the pancreas (Amorim et al., 2020).

4. The role of axial mesoderm in mouse

4.1 Mouse axial mesoderm development

In the mouse, the PMe arises from cells that ingress through the anterior primitive streak in the early-mid streak stage embryo and migrate anteriorly prior to the formation of a morphologically-discrete node. The node forms only in the anterior-most primitive streak of the late streak stage embryo, and cells that transit through it at this point give rise to anterior chordamesoderm. Trunk notochord arises, first from cells that pass through the node of early bud stage embryos, and then from NMPs (Kinder et al., 2001; Tzouanacou, Wegener, Wymeersch, Wilson, & Nicolas, 2009).

As in other vertebrates, graded Nodal activity governs PMe and notochord identities (see Box 2) but additional tissues contribute to PMe specification and maintenance. Intriguingly, a signal(s) from the anterior chordamesoderm reinforces the identity and survival of the mouse PMe at headfold stages. Ablation of the anterior chordamesoderm from late-bud stage embryos (either physically or genetically via *Zic2* null mutation) leads to loss of PMe identity (Barratt et al., 2022; Camus et al., 2000; Warr et al., 2008). Conversely ablation of the late-bud stage PMe leads to regeneration of PMe-specific gene expression (Camus et al., 2000). Furthermore, when transplanted to an ectopic position, the late-bud-PMe loses PMe-specific gene expression, which can be rescued through a transplant of the anterior chordamesoderm (Camus et al., 2000). Thus, anterior chordamesoderm promotes PMe identity and survival, pointing again to the sophisticated interactions that exist between post-involuted PMe and notochord.

4.2 Mouse prechordal mesoderm/mesendoderm and notochord act as local organizers

In mouse, embryonic organizing activity—the ability to organize a complete body axis—is dispersed in space and time, and 'head' organizing activity does not precisely correlate with the PMe, or PMe precursors. When grafted into a mouse embryo, the mouse node (containing notochord precursors) or anterior primitive streak from earlier streak stages (containing PMe precursors) induce only a partial axis, made up of post-occipital regions and lacking a head and forebrain (Beddington, 1994; Martinez Arias & Steventon, 2018; Tam & Beddington, 1992). Instead, embryonic organizing activity appears to require signals from multiple tissues, including PMe and notochord precursors and the anterior visceral endoderm (AVE), an extra-embryonic tissue that is established in distal parts of the pre-gastrula stage embryo, prior to formation of the primitive streak (Kinder et al., 2001). Therefore, in mouse, axial mesoderm is generated from cells that possess only partial embryonic organizing activity, and it is generally accepted that some aspect of 'head' organizing activity is encoded in endodermal tissues (AVE and anterior endoderm) that migrate anteriorly ahead of the PMe (Beddington & Robertson, 1998; Kinder et al., 2001; Martinez Arias & Steventon, 2018). Thus in mouse, factors that direct 'head organization are likely to be distributed across the AVE and anterior endoderm, as well as PMe precursors. (See also Stern & Downs, 2012 and accompanying chapter by Stern (2024) for a discussion of the potential role of the AVE).

Nonetheless, as in chick, mouse PMe regionalizes neural tissue along both the A-P and D-V axes. Ex vivo recombinates show that PMe/anterior chordamesoderm can induce the midbrain marker, Engrailed, in prospective hindbrain tissue (Ang & Rossant, 1993). Likewise, the PMe confers D-V regional identity on forebrain tissue. Manipulations that lead to the loss of the PMe result in D-V forebrain patterning defects, including an expansion of dorsal markers and a loss or reduction of hypothalamic markers (Barratt et al., 2022; Camus et al., 2000; Shimamura & Rubenstein, 1997; Warr et al., 2008). In such embryos, the eye field frequently fails to split, leading to cyclopia. Ex vivo studies likewise show a requirement for the PMe in induction of the hypothalamic markers, *Nkx2–1* and *Shh* (Shimamura & Rubenstein, 1997), although, as in chick, the requirement for the PMe is extremely transient. Finally, grafts of the PMe to diencephalic/prethalamic regions leads to a repression of dorsal markers. Mouse genetic studies have defined transcription factors and signaling molecules that are necessary for forebrain development and are expressed in the PMe. Many of these are the same highly conserved genes shown to direct head development in Xenopus and zebrafish, including *Gsc, Otx2, Lim1, Hesx1*, as well as BMP and Wnt antagonists. $Gsc^{-/-}$ $HNF3\beta^{+/-}$ double mutants, as well as mutants in *Otx2, Lim1, Hesx1* are impaired in anterior neural patterning (Barratt et al., 2022). In the mouse, embryos lacking *Dkk1* activity display ectopic and elevated Wnt signaling activity in precursor tissues of the embryonic head and fail to form brain and head structures (Lewis et al., 2008; Mukhopadhyay et al., 2001). But since *Dkk1* is expressed in additional tissues that contribute to forebrain development, this does not directly prove a role for the PMe. However, forebrain specification is also disrupted in the mouse *nearly headless* (*nehe*) mutant, and in such embryos, the PMe fails to develop, while both the AVE and the anterior endoderm develop normally (Zhou & Anderson, 2010). *Nehe* encodes Lipoic acid synthase, an enzyme required for oxidative metabolism, suggesting that high levels of energy production are required for the specialized morphogenetic movements that generate and characterise the PMe. Finally, as in chick, mouse notochord regionalises both neural tissue and somites along the D-V axes (Chamberlain, Jeong, Guo, Allen, & McMahon, 2008; Chiang et al., 1996; Corallo et al., 2015; Dietrich, Schubert, Gruss, & Lumsden, 1999; Shimamura & Rubenstein, 1997), and is required to pattern the intervertebral discs and the vertebral column (Choi, Lee, & Harfe, 2012).

4.3 Evolutionarily-conserved secreted factors mediate PMe and notochord activities

Molecular studies have identified evolutionarily-conserved secreted factors that induce neural tissue and pattern the neural plate and neural tube along its A-P and D-V axes. While the mechanisms through which neural tissue is generated are different at different axial levels (Henrique et al., 2015; Martinez Arias & Steventon, 2018; Masak & Davidson, 2023), there is widespread agreement that BMP antagonism and FGF signaling are key, and together, induce and maintain neural tissue (Bier & De Robertis, 2015; De Robertis & Kuroda, 2004; De Robertis, 2001, 2009; Harland & Gerhart, 1997; Hemmati-Brivanlou & Melton, 1997; Levine & Brivanlou, 2007; Popov et al., 2017; Stern, 2005; see also accompanying chapter by Slack (Slack, 2024)). A gradient of Wnt signaling (high posterior) provides positional information to the neural plate/neural tube along its A-P axis, and by counteracting Wnt signals, Wnt antagonists direct or maintain anterior identity in neural tissue (Andoniadou & Martinez-Barbera, 2013; Hashimoto et al.,. 2000; Levine & Brivanlou, 2007; Leyns et al., 1997; Slack, 2024; Stern, 2005; Wilson & Houart, 2004; Zhang et al., 2015). Thus anterior neural tissue depends on the dual inhibition of BMP and Wnt signals. Additional factors, including retinoic acid, operate with Wnt to promote posterior identity (Nolte, De Kumar, & Krumlauf, 2019), and retinoic acid antagonists act with Wnt antagonists to protect and maintain anterior neural tissue (Ribes, Fraulob, Petkovich, & Dollé, 2007). Along its D-V axis, the neural tube is patterned by the ventralizing morphogen,

Sonic hedgehog (Shh) (Chang et al., 1994; Chiang et al., 1996; Echelard et al., 1993; Geng et al., 2008; Krauss, Concordet, & Ingham, 1993; Placzek & Briscoe, 2018; Riddle, Johnson, Laufer, & Tabin, 1993; Roelink et al., 1994). In the posterior neural tube, Shh is antagonized by members of the BMP and Wnt families, which act as dorsalizing factors (Sagner & Briscoe, 2019). These factors, together with additional signals, play a critical role in mediating the local organizing activities of the PMe and notochord.

4.4 Prechordal mesoderm/mesendoderm-derived signaling factors

While the exact contribution of the PMe varies from species to species (see above), a common set of evolutionarily-conserved signaling factors contribute to its ability to regionalize tissues and/or maintain regional identity along the A-P and D-V axes. BMP antagonists are transiently expressed in the PMe as it is first generated, although rapidly downregulated (e.g. Dale et al., 1999), and it is likely that they mediate the ability of young head process mesendoderm/PMe to maintain or reinforce neural identity. Wnt antagonists are maintained in the PMe even in the neurula stage embryo (Kim et al., 2022; Niehrs; 1999; Tanaka et al., 2017), and as discussed above, there is a requirement for *Dkk-1* to maintain anterior identity until neurula stages of development (Kim et al., 2009; Shinya et al., 2000). Thus, while other tissues provide Wnt antagonists at earlier pre-gastrula/early gastrula stages, it is highly likely that Wnt antagonists deriving from the PMe regulate ambient Wnt levels to protect cells from the posteriorizing activity of Wnt in the late gastrula and neurula stage embryo, and hence impose and maintain forebrain identities. Likewise retinoic acid clearing enzymes are expressed in the neurula stage PMe in numerous species, and likely create a sink for posteriorizing retinoic acid (Ribes et al., 2007; Simeone et al., 1995).

In terms of D-V patterning, numerous lines of evidence show that the secreted glycoprotein, Shh, derives from the PMe (Ellis et al., 2015; Patten et al., 2003; Placzek & Briscoe, 2018; Riddle et al., 1993; Roelink et al., 1994) and mediates its ability to induce ventral forebrain cell types and impose ventral pattern. Thus embryos that are defective in Shh/Shh signaling components are cyclopic and show D-V patterning defects in the ventral forebrain (Barratt et al., 2022; Chiang et al., 1996; Placzek & Briscoe, 2018). Importantly, mouse embryos in which *Shh* is conditionally deleted in the PMe show cyclopia and defective forebrain D-V patterning (Aoto et al., 2009), while blockade of Shh in the chick PMe eliminates its ability to promote ventral forebrain pattern (Dale et al., 1997). Therefore, while other tissues, including anterior endoderm, anterior chordamesoderm and hypothalamic floor plate, provide additional sources of Shh and likely contribute to forebrain D-V pattern, these studies point to the importance of PMe-derived Shh in either the initiation or the maintenance of forebrain D-V patterning. Intriguingly, studies in mice suggest that Shh deriving from the anterior chordamesoderm may also stabilize the PMe itself, suggesting a role for Shh in both D-V and A-P patterning events (Barratt et al., 2022).

ScRNA-seq analyses in zebrafish point to numerous secreted factors within the PMe, that are expressed in a dynamic manner (Farrell et al., 2018). Potentially, many of these work in conjunction with Shh, Wnt antagonists and RA antagonists in the PMe. The dynamic activity of PMe-derived factors is best understood in the chick. Here, at gastrula stages, the PMe expresses Nodal, which acts cooperatively with Shh to induce hypothalamic floor plate-like cells (Patten et al., 2003; Placzek & Briscoe, 2005). At neurula stages, however, BMPs are upregulated in the PMe (mediated by a signal deriving from anterior endoderm) and consequently, the BMP signaling pathway is activated, and Shh silenced in the PMe (Chinnaiya et al., 2023; Ellis et al., 2015; Vesque et al., 2000). This very dynamic signaling signature is essential for the chick PMe to mediate patterning. Ex vivo and in vivo studies show that, in contrast to their role in the spinal cord, where they dorsalize neural tissue, BMPs now cooperate with Shh to continue to promote the coordinated development of anterior-ventral hypothalamic fates (Chinnaiya et al., 2023; Dale et al., 1997, 1999). Recent work describes that BMPs (potentially deriving from the PMe) promote hypothalamic identity by antagonizing Follistatin, which is widely expressed in the anterior neural plate at gastrula stages, and downregulated as hypothalamic cells are induced and patterned (Kim et al., 2022). Thus, Nodal, BMP and Shh deriving from the PMe may work cooperatively to confer ventral identity. At the same time, BMPs deriving from the PMe also confer anterior identity on neural tissue: exposure of prospective hindbrain ventral cells to BMPs directs them to an anterior hypothalamic identity, coexpressing Shh and Nkx2.1 (Dale et al., 1999). The mechanism remains unclear, but BMPs repress expression of Nkx6.1, so potentially BMPs alleviate a posteriorizing 'repression' (Qiu et al., 1998). Thus, BMPs deriving from the PMe in the neurula stage embryo coordinate D-V and A-P patterning. Other signaling factors likewise may contribute to the ability of the PMe to confer D-V and A-P regional identity, including members of the Fgf family (Chinnaiya et al., 2023; Fu, Towers, & Placzek, 2017; Kim et al., 2022). Intriguingly, in zebrafish, chordin promotes the expression of *dkk1* in the gastrula stage embryo, but Fgf signaling maintains the expression of *dkk1* in the neurula stage embryo (Tanaka et al., 2017). Potentially, therefore, Fgfs reinforce anterior identity by antagonising Wnt signaling.

The PMe, then, has a very distinctive signaling signature to the notochord, but an unanswered question is how these factors operate, mechanistically. Classic studies favor the idea that PMe-derived signals provide positional information to neural tissue, that then expands isotropically (Puelles, 2021). Recent work, however, reveals a massive expansion of the hypothalamus at neurula stages of development, accompanied by anisotropic growth (Chinnaiya et al., 2023; Fu et al., 2017; Kim et al., 2022). Future studies are required to better understand how the PMe may provide instructions to coordinate fate, pattern and proliferation in the developing ventral brain.

Additionally, as discussed above, signals deriving from the young head process mesendoderm and then anterior notochord/PMe promote morphogenesis through cell convergence-migration, through an unknown mechanism (Yoshihi et al., 2022). This activity provides a rationale for how young head process mesendoderm, and then anterior notochord/PMe might consolidate neural identity. As young head process mesendoderm develops it expresses many of the secreted factors that characterize Hensen's node and that contribute to neural induction and neural stability, including BMP antagonists and Fgfs. However, as it resolves into anterior chordamesoderm and PMe, BMP antagonists are retained only in the former, while Wnt antagonists, which promote anterior identity (see below) are expressed at high levels in the latter (eg Chapman et al., 2004; Chapman, Schubert, Schoenwolf, & Lumsden, 2002; Dale et al., 1999; Kim et al., 2022). Potentially, then, neural identity is consolidated as cells migrate over prospective chordamesoderm, and then are anteriorized to a forebrain identity as they migrate over the PMe. In this scenario, cells are induced to anterior neural identity through their sequential exposure to chordamesoderm and then PMe.

4.5 Notochord-derived signaling factors

In all species examined, the secreted glycoprotein, *Shh* is expressed in the notochord and then in the floor plate (Chang et al., 1994; Echelard et al., 1993; Krauss et al., 1993; Riddle et al., 1993; Roelink et al., 1994), and plays a central role in their ability to promote or maintain local organization along the D-V axis (Placzek & Briscoe, 2018). Ex vivo studies, exposing naïve neural explants to purified Shh showed that Shh can induce each of the cell types normally found in the posterior ventral neural tube (Martí, Bumcrot, Takada, & McMahon, 1995; Roelink et al., 1995), and induces different cell types at different concentrations, higher concentrations inducing floor plate, and lower concentrations inducing motor neurons (Ericson et al., 1997; Roelink et al., 1995), suggesting that Shh acts as a morphogen. Through genetic deletion or biochemical blockade, the requirement for Shh was demonstrated in both mice and chick: embryos developed without floor plate and ventral neurons, mimicking the phenotypes of notochord-ablated chick embryos (Chiang et al., 1996; Ericson, Morton, Kawakami, Roelink, & Jessell, 1996). Subsequent studies, exposing prospective hindbrain, midbrain and forebrain and neural explants to Shh showed similarly that Shh can induce and pattern the D-V axis along the entire A-P extent of the neural plate. In each case the type of neuron that is induced reflects an A-P tissue competence: thus, for instance, dopaminergic neurons are induced in the midbrain, while serotonergic neurons are induced in the hindbrain (Ericson et al., 1995). Recent work suggests that Shh deriving from the notochord may play a similar role on humans. The development of three-dimensional stem cell differentiation models, termed gastruloids, has made it possible to compare developmental events between tissues in mouse and human (Moris, Martinez Arias, & Steventon, 2020). These in vitro cultured cells break symmetry, differentiate into all three germ layers, and elongate. giving rise to structures that resemble post-occipital parts of the A-P axis (Beccari et al., 2018; Moris, Anlas et al., 2020; Turner et al., 2017). Neural tissues that are generated within gastruloids that do not express Shh are disorganized; by contrast, under appropriate conditions, gastruloids can be generated that express Shh in cells that resemble a notochord, and in these, overlying neural tissue shows classic D-V patterning. Therefore, potentially, Shh deriving from the notochord acts as an organizer of D-V neural pattern in all amniotes, including humans.

In vivo, however, the organizing activity of the notochord is unlikely to be mediated solely by Shh. The early neural plate and neural tube are exposed to BMP signals, deriving from the epidermal ectoderm and roof plate, and BMP signaling is active within neural tube cells along the D-V axis. In the posterior neuraxis, BMPs control dorsal neural tube cell fate and modify the response to Shh signaling, by inhibiting the specification of a ventral cell fate (Lee & Jessell, 1998; Liem, Tremml, Roelink, & Jessell, 1994; Tozer, Le Dréau, Marti, & Briscoe, 2013). However, in all species, the notochord expresses BMP antagonists, including *Chordin*, reflecting its development from 'embryonic organizer' cells. Through their action in inhibiting BMP, or the BMP signaling pathway, these may shape a BMP morphogen-like gradient. At present it is not clear whether BMP antagonists simply create a permissive environment for the Shh-mediated induction of the floor plate, or whether they play a wider role, effectively creating a pre-pattern along the D-V axis. Strikingly, while some studies have shown that exposure of the dorsal neural tube to Shh induces floor plate and motor neurons (Echelard et al., 1993; Krauss et al., 1993; Roelink et al., 1994), another study has found that Shh alone is unable to recapitulate the activity of transplanted notochord. Instead, only the combined activities of Shh and chordin are able to induce ventral fate and pattern (Patten & Placzek, 2002). Additional studies, then, including those that combine grafting experiments with transcriptomic analyses are needed to better understand the manner in which the graded activity of Shh signaling is interpreted against the background of existing pattern, established by earlier acting gene regulatory networks in particular species (Balaskas et al., 2012; Blitz & Cho, 2021; Charney et al., 2017; Kakebeen et al., 2021; Kiecker et al., 2016; Martinez Arias & Steventon, 2018; Popov et al., 2017).

The same holds true for the ability of the notochord to induce fates and pattern in somites. In vitro and in vivo studies indicate that a Shh gradient works in a morphogen-like fashion to differentially pattern D-V somitic fates, but acts in conjunction with other signals. Very low Shh levels, combined with Wnt signaling activity, elicit the maintenance of dermomyotomal gene expression; slightly higher Shh levels results in the loss of dermomyotomal marker gene expression and activation of myogenic differentiation programmes; finally, high Shh levels cause the loss of myotomal markers and the activation of sclerotomal gene expression (Cairns, Sato, Lee, Lassar, & Zeng, 2008). (Note, some degree of Shh activity is required for the differentiation of dermomyotomal cells into differentiated myocytes, and in the absence of Shh signaling, dermomyotomal cells increase in number, but fail to activate the myogenic program (Feng, Adiarte, & Devoto, 2006; Hammond et al., 2007)). The BMP antagonist, Noggin, deriving from the notochord, cooperates with Shh to mediate sclerotome patterning: Noggin inhibits the repressive activity of BMP signals, thus allowing the Shh-mediated induction of *Pax1* in the somitic mesenchyme and the sclerotomal cell growth and differentiation into cartilage (McMahon et al., 1998). Again, future studies are needed to better understand the existing gene regulatory network that exists in somites prior to the action of Shh, and to establish potential differences in this gene regulatory network across species.

Finally, as in the PMe, additional signals deriving from the notochord provide further cues: for instance, activin-βB and fibroblast growth factor 2 expressed by the notochord repress endodermal Shh, thereby permitting pancreatic dorsal bud development (Corallo et al., 2015; Hebrok, Kim, & Melton, 1998).

5. Summary

The last century has seen enormous advances in our understanding of how the body plan is established, and how neural tissue is induced and patterned. Contemporary studies show that neural induction and patterning occur gradually over time, and are initiated in the pre-gastrula stage embryo, prior to the appearance of midline axial mesoderm. At gastrula and neurula stage embryos, however, midline axial mesoderm plays a critical role, the PMe, in particular, providing crucial signals to induce and maintain anterior identity. The PMe operates at a time of dramatic morphogenetic changes, and so ensures resilience to head and forebrain development. At the same time, axial mesoderm more generally operates to induce and direct pattern widely along the D-V axis. Despite these advances, numerous questions remain unanswered, in particular the role of anterior axial mesoderm in promoting proliferation and morphogenesis. Future studies are required to understand how axial mesoderm coordinates fate, pattern, proliferation and morphogenesis to robustly build the body.

References

Amorim, J. P., Gali-Macedo, A., Marcelino, H., Bordeira-Carriço, R., Naranjo, S., Rivero-Gil, S., et al. (2020). A conserved notochord enhancer controls pancreas development in vertebrates. *Cell Reports, 32*, 107862.

Anderson, C., & Stern, C. D. (2016). Organizers in development. *Current Topics in Developmental Biology, 117*, 435–454.

Andoniadou, C. L., & Martinez-Barbera, J. P. (2013). Developmental mechanisms directing early anterior forebrain specification in vertebrates. *Cellular and Molecular Life Sciences, 70*, 3739–3752.

Ang, S. L., & Rossant, J. (1993). Anterior mesendoderm induces mouse Engrailed genes in explant cultures. *Development (Cambridge, England), 118*, 139–149.

Aoto, K., Shikata, Y., Imai, H., Matsumaru, D., Tokunaga, T., Shioda, S., et al. (2009). Mouse Shh is required for prechordal plate maintenance during brain and craniofacial morphogenesis. *Developmental Biology, 327*, 106–120.

Artinger, K. B., & Bronner-Fraser, M. (1993). Delayed formation of the floor plate after ablation of the avian notochord. *Neuron, 11*, 1147–1161.

Artinger, M., Blitz, I., Inoue, K., Tran, U., & Cho, K. W. (1997). Interaction of goosecoid and brachyury in Xenopus mesoderm patterning. *Mechanisms of Development, 65*(1-2), 187–196.

Balaskas, N., Ribeiro, A., Panovska, J., Dessaud, E., Sasai, N., Page, K. M., et al. (2012). Gene regulatory logic for reading the Sonic Hedgehog signaling gradient in the vertebrate neural tube. *Cell, 148*(1-2), 273–284.

Barratt, K. S., Drover, K. A., Thomas, Z. M., & Arkell, R. M. (2022). Patterning of the antero-ventral mammalian brain: Lessons from holoprosencephaly comparative biology in man and mouse. *WIREs Mechanisms of Disease, 14*(4), e1552.

Beccari, L., Moris, N., Girgin, M., Turner, D. A., Baillie-Johnson, P., Cossy, A. C., ... Arias, A. M. (2018). Multi-axial self-organization properties of mouse embryonic stem cells into gastruloids. *Nature, 562*, 272–276. https://doi.org/10.1038/s41586-018-0578-0.

Beddington, R. S. (1994). Induction of a second neural axis by the mouse node. *Development (Cambridge, England), 120*(3), 613–620.

Beddington, R. S., & Robertson, E. J. (1998). Anterior patterning in mouse. *Trends in Genetics, 14*(7), 277–284.

Bier, E., & De Robertis, E. M. (2015). Embryo development. BMP gradients: A paradigm for morphogen-mediated developmental patterning. *Science (New York, N. Y.), 348*(6242), aaa5838.

Blagden, C. S., Currie, P. D., Ingham, P. W., & Hughes, S. M. (1997). Notochord induction of zebrafish slow muscle mediated by Sonic hedgehog. *Genes & Development (Cambridge, England), 11*(17), 2163–2175.

Blitz, I. L., & Cho, K. W. Y. (2021). Control of zygotic genome activation in Xenopus. *Current Topics in Developmental Biology, 145*, 167–204.

Brand-Saberi, B., Ebensperger, C., Wilting, J., Balling, R., & Christ, B. (1993). The ventralizing effect of the notochord on somite differentiation in chick embryos. *Anatomy and Embryology, 188*(3), 239–245.

Burbridge, S., Stewart, I., & Placzek, M. (2016). Development of the neuroendocrine hypothalamus. *Comprehensive Physiology, 6*(2), 623–643.

Cairns, D. M., Sato, M. E., Lee, P. G., Lassar, A. B., & Zeng, L. (2008). A gradient of Shh establishes mutually repressing somitic cell fates induced by Nkx3.2 and Pax3. *Developmental Biology, 323*, 152–165.

Camus, A., Davidson, B. P., Billiards, S., Khoo, P., Rivera-Perez, J. A., Wakamiya, M., ... Behringer, R. R. (2000). The morphogenetic role of midline mesendoderm and ectoderm in the development of the forebrain and the midbrain of the mouse embryo. *Development (Cambridge, England), 127*, 1799–1813.

Caneparo, L., Huang, Y. L., Staudt, N., Tada, M., Ahrendt, R., Kazanskaya, O., ... Houart, C. (2007). Dickkopf-1 regulates gastrulation movements by coordinated modulation of Wnt/beta-catenin and Wnt/PCP activities, through interaction with the Dally-like homolog Knypek. *Genes & Development, 21*(4), 465–480. https://doi.org/10.1101/gad.406007.

Chamberlain, C. E., Jeong, J., Guo, C., Allen, B. L., & McMahon, A. P. (2008). Notochord-derived Shh concentrates in close association with the apically positioned basal body in neural target cells and forms a dynamic gradient during neural patterning. *Development (Cambridge, England), 135*(6), 1097–1106. https://doi.org/10.1242/dev.013086.

Chang, D. T., López, A., von Kessler, D. P., Chiang, C., Simandl, B. K., Zhao, R., ... Beachy, P. A. (1994). Products, genetic linkage and limb patterning activity of a murine hedgehog gene. *Development (Cambridge, England), 120*(11), 3339–3353. https://doi.org/10.1242/dev.120.11.3339.

Chapman, S. C., Schubert, F. R., Schoenwolf, G. C., & Lumsden, A. (2002). Analysis of spatial and temporal gene expression patterns in blastula and gastrula stage chick embryos. *Developmental Biology, 245*(1), 187–199. https://doi.org/10.1006/dbio.2002.0641.

Chapman, S. C., Brown, R., Lees, L., Schoenwolf, G. C., & Lumsden, A. (2004). Expression analysis of chick Wnt and frizzled genes and selected inhibitors in early chick patterning. *Developmental Dynamics: An Official Publication of the American Association of Anatomists, 229*(3), 668–676. https://doi.org/10.1002/dvdy.10491.

Charney, R. M., Paraiso, K. D., Blitz, I. L., & Cho, K. W. Y. (2017). A gene regulatory program controlling early Xenopus mesendoderm formation: Network conservation and motifs. *Seminars in Cell & Developmental Biology, 66*, 12–24. https://doi.org/10.1016/j.semcdb.2017.03.003.

Cheng, T., Xing, Y. Y., Liu, C., Li, Y. F., Huang, Y., Liu, X., ... Xu, P. F. (2023). Nodal coordinates the anterior-posterior patterning of germ layers and induces head formation in zebrafish explants. *Cell Reports, 42*(4), 112351. https://doi.org/10.1016/j.celrep.2023.112351.

Chiang, C., Litingtung, Y., Lee, E., Young, K. E., Corden, J. L., Westphal, H., & Beachy, P. A. (1996). Cyclopia and defective axial patterning in mice lacking Sonic hedgehog gene function. *Nature, 383*(6599), 407–413. https://doi.org/10.1038/383407a0.

Chinnaiya, K., Burbridge, S., Jones, A., Kim, D. W., Place, E., Manning, E., ... Placzek, M. (2023). A neuroepithelial wave of BMP signaling drives anteroposterior specification of the tuberal hypothalamus. *eLife, 12*, e83133. https://doi.org/10.7554/eLife.83133.

Cho, K. W., Blumberg, B., Steinbeisser, H., & de Robertis, E. M. (1991). Molecular nature of Spemann's organizer: The role of the Xenopus homeobox gene goosecoid. *Cell, 67*(6), 1111–1120.

Choi, K. S., Lee, C., & Harfe, B. D. (2012). Sonic hedgehog in the notochord is sufficient for patterning of the intervertebral discs. *Mechanisms of Development, 129*(9-12), 255–262. https://doi.org/10.1016/j.mod.2012.07.003.

Cleaver, O., Seufert, D. W., & Krieg, P. A. (2000). Endoderm patterning by the notochord: Development of the hypochord in Xenopus. *Development (Cambridge, England), 127*, 869–879.

Corallo, D., Trapani, V., & Bonaldo, P. (2015). The notochord: Structure and functions. *Cellular and Molecular Life Sciences, 72*(16), 2989–3008. https://doi.org/10.1007/s00018-015-1897-z.

Dale, L., & Slack, J. M. (1987). Regional specification within the mesoderm of early embryos of Xenopus laevis. *Development (Cambridge, England), 100*(2), 279–295. https://doi.org/10.1242/dev.100.2.279.

Dale, J. K., Vesque, C., Lints, T. J., Sampath, T. K., Furley, A., Dodd, J., & Placzek, M. (1997). Cooperation of BMP7 and SHH in the induction of forebrain ventral midline cells by prechordal mesoderm. *Cell, 90*(2), 257–269.

Dale, K., Sattar, N., Heemskerk, J., Clarke, J. D., Placzek, M., & Dodd, J. (1999). Differential patterning of ventral midline cells by axial mesoderm is regulated by BMP7 and chordin. *Development (Cambridge, England), 126*(2), 397–408. https://doi.org/10.1242/dev.126.2.397.

Darnell, D. K., Schoenwolf, G. C., & Ordahl, C. P. (1992). Changes in dorsoventral but not rostrocaudal regionalization of the chick neural tube in the absence of cranial notochord, as revealed by expression of engrailed-2. *Developmental Dynamics, 193*(4), 389–396. https://doi.org/10.1002/aja.1001930411.

De Robertis, E. M., Blum, M., Niehrs, C., & Steinbeisser, H. (1992). Goosecoid and the organizer. *Developmental Biology Supplement,* 167–171 PMID: 1363720.

De Robertis, E. M., Fainsod, A., Gont, L. K., & Steinbeisser, H. (1994). The evolution of vertebrate gastrulation. *Developmental Biology, Suppl.,* 117–124 PMID: 7579512.

De Robertis, E. M. (2009). Spemann's organizer and the self-regulation of embryonic fields. *Mechanisms of Development, 126*(11–12), 925–941. https://doi.org/10.1016/j.mod.2009.08.004 PMID: 19733655; PMCID: PMC2803698.

De Robertis, E. M., & Kuroda, H. (2004). Dorsal-ventral patterning and neural induction in Xenopus embryos. *Annual Review of Cell and Developmental Biology, 20*, 285–308. https://doi.org/10.1146/annurev.cellbio.20.011403.154124 PMID: 15473842; PMCID: PMC2280069.

Dias, M. S., & Schoenwolf, G. C. (1990). Formation of ectopic neurepithelium in chick blastoderms: Age-related capacities for induction and self-differentiation following transplantation of quail Hensen's nodes. *Anatomical Record, 228*(4), 437–448. https://doi.org/10.1002/ar.1092280410 PMID: 2285160.

Dietrich, S., Schubert, F. R., Gruss, P., & Lumsden, A. (1999). The role of the notochord for epaxial myotome formation in the mouse. *Cellular and Molecular Biology (Noisy-le-grand), 45*(5), 601–616.

Domingo, C., & Keller, R. (1995). Induction of notochord cell intercalation behavior and differentiation by progressive signals in the gastrula of Xenopus laevis. *Development (Cambridge, England), 121*, 3311–3321.

Doniach, T., Phillips, C. R., & Gerhart, J. C. (1992). Planar induction of anteroposterior pattern in the developing central nervous system of Xenopus laevis. *Science (New York, N. Y.), 257*(5069), 542–545. https://doi.org/10.1126/science.1636091 PMID: 1636091.

Dubrulle, J., Jordan, B. M., Akhmetova, L., Farrell, J. A., Kim, S. H., Solnica-Krezel, L., & Schier, A. F. (2015). Response to Nodal morphogen gradient is determined by the kinetics of target gene induction. *eLife, 4*, e05042. https://doi.org/10.7554/eLife.05042 PMID: 25869585; PMCID: PMC4395910.

Echelard, Y., Epstein, D. J., St-Jacques, B., Shen, L., Mohler, J., McMahon, J. A., & McMahon, A. P. (1993). Sonic hedgehog, a member of a family of putative signaling molecules, is implicated in the regulation of CNS polarity. *Cell, 75*(7), 1417–1430. https://doi.org/10.1016/0092-8674(93)90627-3 PMID: 7916661.

Ellis, P. S., Burbridge, S., Soubes, S., Ohyama, K., Ben-Haim, N., Chen, C., ... Placzek, M. (2015). ProNodal acts via FGFR3 to govern duration of Shh expression in the prechordal mesoderm. *Development (Cambridge, England), 142*(22), 3821–3832. https://doi.org/10.1242/dev.119628 PMID: 26417042; PMCID: PMC4712875.

Ericson, J., Muhr, J., Placzek, M., Lints, T., Jessell, T. M., & Edlund, T. (1995). Sonic hedgehog induces the differentiation of ventral forebrain neurons: A common signal for ventral patterning within the neural tube. *Cell, 81*(5), 747–756. https://doi.org/10.1016/0092-8674(95)90536-7 PMID: 7774016.

Ericson, J., Morton, S., Kawakami, A., Roelink, H., & Jessell, T. M. (1996). Two critical periods of Sonic Hedgehog signaling required for the specification of motor neuron identity. *Cell, 87*(4), 661–673. https://doi.org/10.1016/s0092-8674(00)81386-0 PMID: 8929535.

Ericson, J., Rashbass, P., Schedl, A., Brenner-Morton, S., Kawakami, A., van Heyningen, V., & Briscoe, J. (1997). Pax6 controls progenitor cell identity and neuronal fate in response to graded Shh signaling. *Cell, 90*(1), 169–180. https://doi.org/10.1016/s0092-8674(00)80323-2 PMID: 9230312.

Farrell, J. A., Wang, Y., Riesenfeld, S. J., Shekhar, K., Regev, A., & Schier, A. F. (2018). Single-cell reconstruction of developmental trajectories during zebrafish embryogenesis. *Science (New York, N. Y.), 360*(6392), eaar3131. https://doi.org/10.1126/science.aar3131.

Fauny, J. D., Thisse, B., & Thisse, C. (2009). The entire zebrafish blastula-gastrula margin acts as an organizer dependent on the ratio of Nodal to BMP activity. *Development (Cambridge, England), 136*(22), 3811–3819. https://doi.org/10.1242/dev.039693 PMID: 19855023.

Fekany-Lee, K., Gonzalez, E., Miller-Bertoglio, V., & Solnica-Krezel, L. (2000). The homeobox gene bozozok promotes anterior neuroectoderm formation in zebrafish through negative regulation of BMP2/4 and Wnt pathways. *Development (Cambridge, England), 127*(11), 2333–2345.

Feng, X., Adiarte, E. G., & Devoto, S. H. (2006). Hedgehog acts directly on the zebrafish dermomyotome to promote myogenic differentiation. *Developmental Biology, 300*, 736–746.

Foley, A. C., Storey, K. G., & Stern, C. D. (1997). The prechordal region lacks neural inducing ability, but can confer anterior character to more posterior neuroepithelium. *Development (Cambridge, England), 124*(17), 2983–2996.

Fu, T., Towers, M., & Placzek, M. A. (2017). Fgf10+ progenitors give rise to the chick hypothalamus by rostral and caudal growth and differentiation. *Development (Cambridge, England), 144*(18), 3278–3288. https://doi.org/10.1242/dev.153379.

Geng, X., Speirs, C., Lagutin, O., Inbal, A., Liu, W., Solnica-Krezel, L., ... Oliver, G. (2008). Haploinsufficiency of Six3 fails to activate Sonic hedgehog expression in the ventral forebrain and causes holoprosencephaly. *Developmental Cell, 15*(2), 236–247. https://doi.org/10.1016/j.devcel.2008.07.003.

Glinka, A., Wu, W., Delius, H., Monaghan, A. P., Blumenstock, C., & Niehrs, C. (1998). Dickkopf-1 is a member of a new family of secreted proteins and functions in head induction. *Nature, 391*(6665), 357–362. https://doi.org/10.1038/34848.

Goldstein, A. M., & Fishman, M. C. (1998). Notochord regulates cardiac lineage in zebrafish embryos. *Developmental Biology, 201*(2), 247–252. https://doi.org/10.1006/dbio.1998.8976.

Gleiberman, A. S., Fedtsova, N. G., & Rosenfeld, M. G. (1999). Tissue interactions in the induction of anterior pituitary: Role of the ventral diencephalon, mesenchyme, and notochord. *Developmental Biology, 213*(2), 340–353. https://doi.org/10.1006/dbio.1999.9386.

Grinblat, Y., Gamse, J., Patel, M., & Sive, H. (1998). Determination of the zebrafish forebrain: Induction and patterning. *Development (Cambridge, England), 125*(22), 4403–4416. https://doi.org/10.1242/dev.125.22.4403 PMID: 9778500.

Gritsman, K., Talbot, W. S., & Schier, A. F. (2000). Nodal signaling patterns the organizer. *Development (Cambridge, England), 127*(5), 921–932. https://doi.org/10.1242/dev.127.5.921.

Guillot, C., Djeffal, Y., Michaut, A., Rabe, B., & Pourquié, O. (2021). Dynamics of primitive streak regression controls the fate of neuromesodermal progenitors in the chicken embryo. *eLife, 10*, e64819. https://doi.org/10.7554/eLife.64819.

Gurdon, J. B. (1987). Embryonic induction—Molecular prospects. *Development (Cambridge, England), 99*, 2285–2306.

Halpern, M. E., Ho, R. K., Walker, C., & Kimmel, C. B. (1993). Induction of muscle pioneers and floor plate is distinguished by the zebrafish no tail mutation. *Cell, 75*, 99–111.

Halpern, M. E., Thisse, C., Ho, R. K., Thisse, B., Riggleman, B., Trevarrow, B., ... Kimmel, C. B. (1995). Cell-autonomous shift from axial to paraxial mesodermal development in zebrafish floating head mutants. *Development (Cambridge, England), 121*, 4257–4264.

Hammond, C. L., Hinits, Y., Osborn, D. P., Minchin, J. E., Tettamanti, G., & Hughes, S. M. (2007). Signals and myogenic regulatory factors restrict pax3 and pax7 expression to dermomyotome-like tissue in zebrafish. *Developmental Biology, 302*, 504–521.

Harland, R., & Gerhart, J. (1997). Formation and function of Spemann's organizer. *Annual Review of Cell and Developmental Biology, 13*, 611–667. https://doi.org/10.1146/annurev.cellbio.13.1.611.

Hashimoto, H., Itoh, M., Yamanaka, Y., Yamashita, S., Shimizu, T., Solnica-Krezel, L., ... Hirano, T. (2000). Zebrafish Dkk1 functions in forebrain specification and axial mesendoderm formation. *Developmental Biology, 217*(1), 138–152. https://doi.org/10.1006/dbio.1999.9537.

Hebrok, M., Kim, S. K., & Melton, D. A. (1998). Notochord repression of endodermal Sonic hedgehog permits pancreas development. *Genes & Development (Cambridge, England), 12*, 1705–1713.

Hemmati-Brivanlou, A., & Melton, D. (1997). Vertebrate neural induction. *Annual Review of Neuroscience, 20*, 43–60. https://doi.org/10.1146/annurev.neuro.20.1.43.

Henrique, D., Abranches, E., Verrier, L., & Storey, K. G. (2015). Neuromesodermal progenitors and the making of the spinal cord. *Development (Cambridge, England), 142*(17), 2864–2875. https://doi.org/10.1242/dev.119768.

Izpisúa-Belmonte, J. C., De Robertis, E. M., Storey, K. G., & Stern, C. D. (1993). The homeobox gene goosecoid and the origin of organizer cells in the early chick blastoderm. *Cell, 74*(4), 645–659. https://doi.org/10.1016/0092-8674(93)90512-o.

Johnson, R. L., Riddle, R. D., Laufer, E., & Tabin, C. (1994). Sonic hedgehog: A key mediator of anterior-posterior patterning of the limb and dorso-ventral patterning of axial embryonic structures. *Biochemical Society Transactions, 22*(3), 569–574. https://doi.org/10.1042/bst0220569.

Jones, W. D., & Mullins, M. C. (2022). Cell signaling pathways controlling an axis organizing center in the zebrafish. *Current Topics in Developmental Biology, 150*, 149–209. https://doi.org/10.1016/bs.ctdb.2022.03.005.

Kakebeen, A. D., Huebner, R. J., Shindo, A., Kwon, K., Kwon, T., Wills, A. E., & Wallingford, J. B. (2021). A temporally resolved transcriptome for developing "Keller" explants of the *Xenopus laevis* dorsal marginal zone. *Developmental Dynamics, 250*(5), 717–731. https://doi.org/10.1002/dvdy.289.

Keller, R. E. (1975). Vital dye mapping of the gastrula and neurula of *Xenopus laevis*. I. Prospective areas and morphogenetic movements of the superficial layer. *Developmental Biology, 42*(2), 222–241. https://doi.org/10.1016/0012-1606(75)90331-0.

Keller, R. E. (1976). Vital dye mapping of the gastrula and neurula of *Xenopus laevis*. II. Prospective areas and morphogenetic movements of the deep layer. *Developmental Biology, 51*(1), 118–137. https://doi.org/10.1016/0012-1606(76)90127-5.

Keller, R., & Danilchik, M. (1988). Regional expression, pattern and timing of convergence and extension during gastrulation of *Xenopus laevis*. *Development (Cambridge, England), 103*(1), 193–209. https://doi.org/10.1242/dev.103.1.193.

Kiecker, C., & Niehrs, C. (2001). The role of prechordal mesendoderm in neural patterning. *Current Opinion in Neurobiology, 11*(1), 27–33. https://doi.org/10.1016/s0959-4388(00)00170-7.

Kiecker, C., Bates, T., & Bell, E. (2016). Molecular specification of germ layers in vertebrate embryos. *Cell and Molecular Life Sciences, 73*(5), 923–947. https://doi.org/10.1007/s00018-015-2092-y.

Kim, D. W., Place, E., Chinnaiya, K., Manning, E., Sun, C., Dai, W., ... Blackshaw, S. (2022). Single-cell analysis of early chick hypothalamic development reveals that hypothalamic cells are induced from prethalamic-like progenitors. *Cell Reports, 38*(3), 110251. https://doi.org/10.1016/j.celrep.2021.110251.

Kim, J. D., Chun, H. S., Kim, S. H., Kim, H. S., Kim, Y. S., Kim, M. J., ... Huh, T. L. (2009). Normal forebrain development may require continual Wnt antagonism until mid-somitogenesis in zebrafish. *Biochemical and Biophysical Research Communications, 381*(4), 717–721. https://doi.org/10.1016/j.bbrc.2009.02.135.

Kim, S. K., Hebrok, M., & Melton, D. A. (1997). Notochord to endoderm signaling is required for pancreas development. *Development (Cambridge, England), 124*, 4243–4252.

Kimmel, C. B., Warga, R. M., & Schilling, T. F. (1990). Origin and organization of the zebrafish fate map. *Development (Cambridge, England), 108*(4), 581–594. https://doi.org/10.1242/dev.108.4.581.

Kinder, S. J., Tsang, T. E., Wakamiya, M., Sasaki, H., Behringer, R. R., Nagy, A., & Tam, P. P. (2001). The organizer of the mouse gastrula is composed of a dynamic population of progenitor cells for the axial mesoderm. *Development (Cambridge, England), 128*(18), 3623–3634. https://doi.org/10.1242/dev.128.18.3623.

Krauss, S., Concordet, J. P., & Ingham, P. W. (1993). A functionally conserved homolog of the Drosophila segment polarity gene hh is expressed in tissues with polarizing activity in zebrafish embryos. *Cell, 75*(7), 1431–1444. https://doi.org/10.1016/0092-8674(93)90628-4.

Kumar, V., Umair, Z., Lee, U., & Kim, J. (2023). Two homeobox transcription factors, goosecoid and Ventx1.1, oppositely regulate chordin transcription in *Xenopus gastrula* embryos. *Cells, 12*(6), 874. https://doi.org/10.3390/cells12060874.

Kwan, K. M. (2003). Xbra functions as a switch between cell migration and convergent extension in the *Xenopus gastrula*. *Development (Cambridge, England), 130*, 1961–1972.

Lee, K. J., & Jessell, T. M. (1998). The specification of dorsal cell fates in the vertebrate central nervous system. *Annual Review of Neuroscience, 22*, 261–294.

Levine, A. J., & Brivanlou, A. H. (2007). Proposal of a model of mammalian neural induction. *Developmental Biology, 308*(2), 247–256. https://doi.org/10.1016/j.ydbio.2007.05.036.

Lewis, S. L., Khoo, P. L., De Young, R. A., Steiner, K., Wilcock, C., Mukhopadhyay, M., ... Tam, P. P. (2008). Dkk1 and Wnt3 interact to control head morphogenesis in the mouse. *Development (Cambridge, England), 135*(10), 1791–1801. https://doi.org/10.1242/dev.018853.

Leyns, L., Bouwmeester, T., Kim, S. H., Piccolo, S., & De Robertis, E. M. (1997). Frzb-1 is a secreted antagonist of Wnt signaling expressed in the Spemann organizer. *Cell, 88*(6), 747–756. https://doi.org/10.1016/s0092-8674(00)81921-2.

Li, H., Tierney, C., Wen, L., Wu, J. Y., & Rao, Y. (1997). A single morphogenetic field gives rise to two retina primordia under the influence of the prechordal plate. *Development (Cambridge, England), 124*, 603–615.

Liem, K. F., Tremml, G., Roelink, H., & Jessell, T. M. (1994). Dorsal differentiation of neural plate cells induced by BMP-mediated signals from epidermal ectoderm. *Cell, 82*, 969–979.

Mangold, O. (1933). Über die induktionsfähigkeit der verschiedenen bezirke der neurula von Urodelen. *Die Naturwissenschaften, 21*, 761–766.

Marcelle, C., Stark, M. R., & Bronner-Fraser, M. (1997). Coordinate actions of BMPs, Wnts, Shh and noggin mediate patterning of the dorsal somite. *Development (Cambridge, England), 124*(20), 3955–3963. https://doi.org/10.1242/dev.124.20.3955 PMID: 9374393.

Martí, E., Bumcrot, D. A., Takada, R., & McMahon, A. P. (1995). Requirement of 19K form of Sonic hedgehog for induction of distinct ventral cell types in CNS explants. *Nature, 375*(6529), 322–325. https://doi.org/10.1038/375322a0 PMID: 7753196.

Martinez Arias, A., & Steventon, B. (2018). On the nature and function of organizers. *Development (Cambridge, England), 145*(5), dev159525. https://doi.org/10.1242/dev.159525 PMID: 29523654; PMCID: PMC5868996.

Masak, G., & Davidson, L. A. (2023). Constructing the pharyngula: Connecting the primary axial tissues of the head with the posterior axial tissues of the tail. *Cells & Development, 176*, 203866. https://doi.org/10.1016/j.cdev.2023.203866 PMID: 37394035.

McMahon, J. A., Takada, S., Zimmerman, L. B., Fan, C. M., Harland, R. M., & McMahon, A. P. (1998). Noggin-mediated antagonism of BMP signaling is required for growth and patterning of the neural tube and somite. *Genes & Development, 12*, 1438–1452.

Moghadam, K. S., Chen, A., & Heathcote, R. D. (2003). Establishment of a ventral cell fate in the spinal cord. *Developmental Dynamics: An Official Publication of the American Association of Anatomists, 227*(4), 552–562. https://doi.org/10.1002/dvdy.10340 PMID: 12889064.

Moris, N., Martinez Arias, A., & Steventon, B. (2020). Experimental embryology of gastrulation: Pluripotent stem cells as a new model system. *Current Opinion in Genetics and Development, 64*, 78–83. https://doi.org/10.1016/j.gde.2020.05.031.

Moris, N., Anlas, K., van den Brink, S. C., Alemany, A., Schröder, J., Ghimire, S., ... Martinez Arias, A. (2020). An in vitro model of early anteroposterior organization during human development. *Nature, 582*, 410–415. https://doi.org/10.1038/s41586-020-2383-9.

Mukhopadhyay, M., Shtrom, S., Rodriguez-Esteban, C., Chen, L., Tsukui, T., Gomer, L., ... Westphal, H. (2001). Dickkopf1 is required for embryonic head induction and limb morphogenesis in the mouse. *Developmental Cell, 1*(3), 423–434. https://doi.org/10.1016/s1534-5807(01)00041-7 PMID: 11702953.

Nicolet, G. (1971). Avian gastrulation. *Advances in Morphogenesis, 9*, 231–262. https://doi.org/10.1016/b978-0-12-028609-6.50010-8 PMID: 4103617.

Niehrs, C. (1999). Head in the WNT: The molecular nature of Spemann's head organizer. *Trends in Genetics: TIG, 15*(8), 314–319.

Niehrs, C. (2022). The role of Xenopus developmental biology in unraveling Wnt signaling and antero-posterior axis formation. *Developmental Biology, 482,* 1–6. https://doi.org/10.1016/j.ydbio.2021.11.006 PMID: 34818531.

Nieuwkoop, P. D. (1997). Short historical survey of pattern formation in the endo-mesoderm and the neural anlage in the vertebrates: The role of vertical and planar inductive actions. *Cellular and Molecular Life Sciences: CMLS, 53*(4), 305–318. https://doi.org/10.1007/pl00000608 PMID: 9137623.

Nolte, C., De Kumar, B., & Krumlauf, R. (2019). Hox genes: Downstream "effectors" of retinoic acid signaling in vertebrate embryogenesis. *Genesis (New York, N. Y.: 2000), 57*(7-8), e23306. https://doi.org/10.1002/dvg.23306 PMID: 31111645.

Odenthal, J., Haffter, P., Vogelsang, E., Brand, M., van Eeden, F. J., Furutani-Seiki, M., & Nüsslein-Volhard, C. (1996). Mutations affecting the formation of the notochord in the zebrafish, Danio rerio. *Development (Cambridge, England), 123,* 103–115. https://doi.org/10.1242/dev.123.1.103 PMID: 9007233.

Ohyama, K., Ellis, P., Kimura, S., & Placzek, M. (2005). Directed differentiation of neural cells to hypothalamic dopaminergic neurons. *Development (Cambridge, England), 132*(23), 5185–5197. https://doi.org/10.1242/dev.02094.

Patten, I., & Placzek, M. (2002). Opponent activities of Shh and BMP signaling during floor plate induction in vivo. *Current Biology, 12*(1), 47–52. https://doi.org/10.1016/s0960-9822(01)00631-5.

Patten, I., Kulesa, P., Shen, M. M., Fraser, S., & Placzek, M. (2003). Distinct modes of floor plate induction in the chick embryo. *Development (Cambridge, England), 130*(20), 4809–4821. https://doi.org/10.1242/dev.00694.

Pera, E. M., & Kessel, M. (1997). Patterning of the chick forebrain anlage by the prechordal plate. *Development (Cambridge, England), 124,* 4153–4162.

Piccolo, S., Agius, E., Leyns, L., Bhattacharyya, S., Grunz, H., Bouwmeester, T., & De Robertis, E. M. (1999). The head inducer cerberus is a multifunctional antagonist of nodal, BMP, and WNT signals. *Nature, 397*(6721), 707–710.

Placzek, M., Tessier-Lavigne, M., Yamada, T., Jessell, T., & Dodd, J. (1990). Mesodermal control of neural cell identity: Floor plate induction by the notochord. *Science (New York, N. Y.), 250*(4983), 985–988. https://doi.org/10.1126/science.2237443.

Placzek, M., Yamada, T., Tessier-Lavigne, M., Jessell, T., & Dodd, J. (1991). Control of dorsoventral pattern in vertebrate neural development: Induction and polarizing properties of the floor plate. *Developmental Biology, Suppl. 2,* 105–122.

Placzek, M., & Briscoe, J. (2005). The floor plate: Multiple cells, multiple signals. *Nature Reviews. Neuroscience, 6*(3), 230–240. https://doi.org/10.1038/nrn1628.

Placzek, M., & Briscoe, J. (2018). Sonic hedgehog in vertebrate neural tube development. *International Journal of Developmental Biology, 62*(1–2–3), 225–234. https://doi.org/10.1387/ijdb.170293jb.

Popov, I. K., Kwon, T., Crossman, D. K., Crowley, M. R., Wallingford, J. B., & Chang, C. (2017). Identification of new regulators of embryonic patterning and morphogenesis in *Xenopus gastrulae* by RNA sequencing. *Developmental Biology, 426*(2), 429–441. https://doi.org/10.1016/j.ydbio.2016.05.014.

Pourquié, O., Coltey, M., Teillet, M. A., Ordahl, C., & Le Douarin, N. M. (1993). Control of dorsoventral patterning of somitic derivatives by notochord and floor plate. *Proceedings of the National Academy of Sciences of the United States of America, 90*(11), 5242–5246. https://doi.org/10.1073/pnas.90.11.5242.

Pownall, M. E., Strunk, K. E., & Emerson, C. P., Jr. (1996). Notochord signals control the transcriptional cascade of myogenic bHLH genes in somites of quail embryos. *Development (Cambridge, England), 122*(5), 1475–1488. https://doi.org/10.1242/dev.122.5.1475.

Puelles, L. (2021). Recollections on the origins and development of the prosomeric model. *Frontiers in Neuroanatomy, 15*, 787913. https://doi.org/10.3389/fnana.2021.787913.

Qiu, M., Shimamura, K., Sussel, L., Chen, S., & Rubenstein, J. L. (1998). Control of anteroposterior and dorsoventral domains of Nkx-6.1 gene expression relative to other Nkx genes during vertebrate CNS development. *Mechanisms of Development, 72*(1-2), 77–88. https://doi.org/10.1016/s0925-4773(98)00018-5.

Ribes, V., Fraulob, V., Petkovich, M., & Dollé, P. (2007). The oxidizing enzyme CYP26a1 tightly regulates the availability of retinoic acid in the gastrulating mouse embryo to ensure proper head development and vasculogenesis. *Developmental Dynamics, 236*(3), 644–653. https://doi.org/10.1002/dvdy.21057.

Riddle, R. D., Johnson, R. L., Laufer, E., & Tabin, C. (1993). Sonic hedgehog mediates the polarizing activity of the ZPA. *Cell, 75*(7), 1401–1416. https://doi.org/10.1016/0092-8674(93)90626-2 PMID: 8269518.

Robertson, E. J. (2014). Dose-dependent Nodal/Smad signals pattern the early mouse embryo. *Seminars in Cell & Developmental Biology, 32*, 73–79. https://doi.org/10.1016/j.semcdb.2014.03.028 PMID: 24704361.

Roelink, H., Augsburger, A., Heemskerk, J., Korzh, V., Norlin, S., Ruiz i Altaba, A., ... Jessell, T. M. (1994). Floor plate and motor neuron induction by vhh-1, a vertebrate homolog of hedgehog expressed by the notochord. *Cell, 76*(4), 761–775. https://doi.org/10.1016/0092-8674(94)90514-2 PMID: 8124714.

Roelink, H., Porter, J. A., Chiang, C., Tanabe, Y., Chang, D. T., Beachy, P. A., & Jessell, T. M. (1995). Floor plate and motor neuron induction by different concentrations of the amino-terminal cleavage product of sonic hedgehog autoproteolysis. *Cell, 81*(3), 445–455. https://doi.org/10.1016/0092-8674(95)90397-6 PMID: 7736596.

Rowan, A. M., Stern, C. D., & Storey, K. G. (1999). Axial mesendoderm refines rostrocaudal pattern in the chick nervous system. *Development (Cambridge, England), 126*(13), 2921–2934. https://doi.org/10.1242/dev.126.13.2921 PMID: 10357936.

Ruiz i Altaba, A. (1990). Neural expression of the Xenopus homeobox gene Xhox3: Evidence for a patterning neural signal that spreads through the ectoderm. *Development (Cambridge, England), 108*(4), 595–604. https://doi.org/10.1242/dev.108.4.595 PMID: 1974841.

Sagner, A., & Briscoe, J. (2019). Establishing neuronal diversity in the spinal cord: A time and a place. *Development (Cambridge, England), 146*(22), dev182154. https://doi.org/10.1242/dev.182154 PMID: 31767567.

Sako, K., Pradhan, S. J., Barone, V., Inglés-Prieto, Á., Müller, P., Ruprecht, V., ... Heisenberg, C. P. (2016). Optogenetic control of nodal signaling reveals a temporal pattern of nodal signaling regulating cell fate specification during gastrulation. *Cell Reports, 16*(3), 866–877. https://doi.org/10.1016/j.celrep.2016.06.036 PMID: 27396324.

Sander, K., & Faessler, P. E. (2001). Introducing the Spemann-Mangold organizer: Experiments and insights that generated a key concept in developmental biology. *International Journal of Developmental Biology, 45*(1), 1–11.

Saúde, L., Woolley, K., Martin, P., Driever, W., & Stemple, D. L. (2000). Axis-inducing activities and cell fates of the zebrafish organizer. *Development (Cambridge, England), 127*, 3407–3417.

Schier, A. F., Neuhauss, S. C., Helde, K. A., Talbot, W. S., & Driever, W. (1997). The one-eyed pinhead gene functions in mesoderm and endoderm formation in zebrafish and interacts with no tail. *Development (Cambridge, England), 124*(2), 327–342. https://doi.org/10.1242/dev.124.2.327.

Schier, A. F., & Shen, M. M. (2000). Nodal signaling in vertebrate development. *Nature, 403*(6768), 385–389. https://doi.org/10.1038/35000126.

Schneider, V. A., & Mercola, M. (1999). Spatially distinct head and heart inducers within the Xenopus organizer region. *Current Biology, 9*(15), 800–809. https://doi.org/10.1016/s0960-9822(99)80363-7.

Schoenwolf, G. C., & Sheard, P. (1990). Fate mapping the avian epiblast with focal injections of a fluorescent-histochemical marker: Ectodermal derivatives. *Journal of Experimental Zoology, 255*(3), 323–339. https://doi.org/10.1002/jez.1402550309.

Schulte-Merker, S., Hammerschmidt, M., Beuchle, D., Cho, K. W., De Robertis, E. M., & Nüsslein-Volhard, C. (1994). Expression of zebrafish goosecoid and no tail gene products in wild-type and mutant no tail embryos. *Development (Cambridge, England), 120*(4), 843–852. https://doi.org/10.1242/dev.120.4.843.

Selleck, M. A., & Stern, C. D. (1991). Fate mapping and cell lineage analysis of Hensen's node in the chick embryo. *Development (Cambridge, England), 112*(2), 615–626. https://doi.org/10.1242/dev.112.2.615 PMID: 1794328.

Serrallach, B. L., Rauch, R., Lyons, S. K., & Huisman, T. A. G. M. (2022). Duplication of the pituitary gland: CT, MRI and DTI findings and updated review of the literature. *Brain Sciences, 12*(5), 574. https://doi.org/10.3390/brainsci12050574.

Seifert, R., Jacob, M., & Jacob, H. J. (1993). The avian prechordal head region: A morphological study. *Journal of Anatomy, 183*(Pt 1), 75–89. PMID: 8270478; PMCID: PMC1259855.

Shimamura, K., & Rubenstein, J. L. (1997). Inductive interactions direct early regionalization of the mouse forebrain. *Development (Cambridge, England), 124*(14), 2709–2718. https://doi.org/10.1242/dev.124.14.2709.

Shih, J., & Keller, R. (1992). The epithelium of the dorsal marginal zone of Xenopus has organizer properties. *Development (Cambridge, England), 116*(4), 887–899. https://doi.org/10.1242/dev.116.4.887.

Simeone, A., Avantaggiato, V., Moroni, M. C., Mavilio, F., Arra, C., Cotelli, F., ... Acampora, D. (1995). Retinoic acid induces stage-specific antero-posterior transformation of rostral central nervous system. *Mechanisms of Development, 51*(1), 83–98. https://doi.org/10.1016/0925-4773(95)96241-m.

Shinya, M., Eschbach, C., Clark, M., Lehrach, H., & Furutani-Seiki, M. (2000). Zebrafish Dkk1, induced by the pre-MBT Wnt signaling, is secreted from the prechordal plate and patterns the anterior neural plate. *Mechanisms of Development, 98*(1-2), 3–17. https://doi.org/10.1016/s0925-4773(00)00433-0.

Slack, J. (2024). The organizer: What it meant, and still means, to developmental biology. *Curr. Top. Dev. Biol. 157*, 1–42.

Soh, G. H., Pomreinke, A. P., & Müller, P. (2020). Integration of nodal and BMP signaling by mutual signaling effector antagonism. *Cell Reports, 31*(1), 107487. https://doi.org/10.1016/j.celrep.2020.03.051.

Sokol, S. (2000). A role for Wnts in morphogenesis and tissue polarity. *Nature Cell Biology, 2*, E124–E125.

Spemann, H. (1929). Über den Anteil von Organisator u. Wirtskeim am Zustandekommen der Induktion. *Die Naturwissenschaften, 17*, 287–289.

Spemann, H. (1931a). Das Verhalten von Organisatoren nach Zerstörung ihrer Struktur. *Verhandlungen der Deutschen Zoologischen Gesellschaft, 1931*, 129–132.

Spemann, H. (1931b). Über den Anteil von Implantat und Wirskeim an der Orientierung und Beschaffenheit der induzierten Embryonalanlage. *Wilhelm Roux' Archiv für Entwicklungsmechanik der Organismen, 123*, 390–517.

Spemann, H., & Mangold, H. (1924). Über Induktion von Embryonalanlagen durch Implantation artfremder Organisatoren. *Wilhelm Roux' Archiv fur Entwicklungsmechanik der Organismen, 100*, 599–638.

Stern, C. D. (2005). Neural induction: Old problem, new findings, yet more questions. *Development (Cambridge, England), 132*(9), 2007–2021. https://doi.org/10.1242/dev.01794.

Stern, C. D (2024). The organizer and neural induction in birds and mammals. *Curr. Top. Dev. Biol. 157*, 43–66.

Stern, C. D., & Downs, K. M. (2012). The hypoblast (visceral endoderm): An evo-devo perspective. *Development (Cambridge, England), 139*(6), 1059–1069. https://doi.org/10.1242/dev.070730.

Storey, K. G., Crossley, J. M., De Robertis, E. M., Norris, W. E., & Stern, C. D. (1992). Neural induction and regionalisation in the chick embryo. *Development (Cambridge, England), 114*(3), 729–741. https://doi.org/10.1242/dev.114.3.729.

Stower, M. J., & Srinivas, S. (2018). The head's tale: Anterior-posterior axis formation in the mouse embryo. *Current Topics in Developmental Biology, 128*, 365–390. https://doi.org/10.1016/bs.ctdb.2017.11.003.

Streit, A., Lee, K. J., Woo, I., Roberts, C., Jessell, T. M., Stern, C. D., & Driever, W. (1998). Chordin regulates primitive streak development and the stability of induced neural cells, but is not sufficient for neural induction in the chick embryo. *Development (Cambridge, England), 125*(3), 507–519. https://doi.org/10.1242/dev.125.3.507.

Talbot, W. S., Trevarrow, B., Halpern, M. E., Melby, A. E., Farr, G., Postlethwait, J. H., ... Kimelman, D. (1995). A homeobox gene essential for zebrafish notochord development. *Nature, 378*, 150–157.

Tam, P. P., & Beddington, R. S. (1992). Establishment and organization of germ layers in the gastrulating mouse embryo. *Ciba Foundation Symposium, 165*, 27–41. https://doi.org/10.1002/9780470514221.ch3.

Tanaka, S., Hosokawa, H., Weinberg, E. S., & Maegawa, S. (2017). Chordin and dickkopf-1b are essential for the formation of head structures through activation of the FGF signaling pathway in zebrafish. *Developmental Biology, 424*(2), 189–197. https://doi.org/10.1016/j.ydbio.2017.02.018.

Thisse, B., & Thisse, C. (2015). Formation of the vertebrate embryo: Moving beyond the Spemann organizer. *Seminars in Cell & Developmental Biology, 42*, 94–102. https://doi.org/10.1016/j.semcdb.2015.05.007.

Tozer, S., Le Dréau, G., Marti, E., & Briscoe, J. (2013). Temporal control of BMP signaling determines neuronal subtype identity in the dorsal neural tube. *Development (Cambridge, England), 140*, 1467–1474. https://doi.org/10.1242/dev.090118.

Trevers, K. E., Prajapati, R. S., Hintze, M., Stower, M. J., Strobl, A. C., Tambalo, M., ... Streit, A. (2018). Neural induction by the node and placode induction by head mesoderm share an initial state resembling neural plate border and ES cells. *Proceedings of the National Academy of Sciences of the United States of America, 115*(2), 355–360. https://doi.org/10.1073/pnas.1719674115.

Trevers, K. E., Lu, H. C., Yang, Y., Thiery, A. P., Strobl, A. C., Anderson, C., ... Stern, C. D. (2023). A gene regulatory network for neural induction. *eLife, 12*, e73189. https://doi.org/10.7554/eLife.73189.

Tucker, J. A., Mintzer, K. A., & Mullins, M. C. (2008). The BMP signaling gradient patterns dorsoventral tissues in a temporally progressive manner along the anteroposterior axis. *Developmental Cell, 14*(1), 108–119. https://doi.org/10.1016/j.devcel.2007.11.004.

Turner, D. A., et al. (2017). Anteroposterior polarity and elongation in the absence of extra-embryonic tissues and of spatially localised signaling in gastruloids: Mammalian embryonic organoids. *Development (Cambridge, England), 144*, 3894–3906. https://doi.org/10.1242/dev.150391.

Tzouanacou, E., Wegener, A., Wymeersch, F. J., Wilson, V., & Nicolas, J. F. (2009). Redefining the progression of lineage segregations during mammalian embryogenesis by clonal analysis. *Developmental Cell, 17*(3), 365–376. https://doi.org/10.1016/j.devcel.2009.08.002.

Ulmer, B., Tingler, M., Kurz, S., Maerker, M., Andre, P., Mönch, D., ... Blum, M. (2017). A novel role of the organizer gene Goosecoid as an inhibitor of Wnt/PCP-mediated convergent extension in Xenopus and mouse. *Scientific Reports, 7*, 43010. https://doi.org/10.1038/srep43010.

Umair, Z., Kumar, V., Goutam, R. S., Kumar, S., Lee, U., & Kim, J. (2021). Goosecoid controls neuroectoderm specification via dual circuits of direct repression and indirect stimulation in Xenopus embryos. *Molecular Cells, 44*(10), 723–735. https://doi.org/10.14348/molcells.2021.0055 PMID: 34711690; PMCID: PMC8560583.

van Straaten, H. W., Hekking, J. W., Wiertz-Hoessels, E. J., Thors, F., & Drukker, J. (1988). Effect of the notochord on the diffcrentiation of a floor plate area in the neural tube of the chick embryo. *Anatomy and Embryology (Berlin), 177*(4), 317–324. https://doi.org/10.1007/BF00315839 PMID: 3354847.

Vesque, C., Ellis, S., Lee, A., Szabo, M., Thomas, P., Beddington, R., & Placzek, M. (2000). Development of chick axial mesoderm: Specification of prechordal mesoderm by anterior endoderm-derived TGFbeta family signaling. *Development (Cambridge, England), 127*(13), 2795–2809. https://doi.org/10.1242/dev.127.13.2795 PMID: 10851126.

Waddington, C. H. (1933). Induction by the primitive streak and its derivatives in the chick. *Journal of Experimental Biology, 10*, 38–46.

Waddington, C. H., & Schmidt, G. A. (1933). Induction by heteroplastic grafts of the primitive streak in birds. *Wilhelm Roux' Archiv für Entwicklungsmechanik der Organismen, 128*, 522–563.

Waddington, C. H. (1934). Experiments on embryonic induction. III. A note on inductions by the chick primitive streak transplanted to the rabbit embryo. *Journal of Experimental Biology, 11*, 224–226.

Waddington, C. H. (1937). Experiments on determination in the rabbit embryo. *Archives of Biology, 48*, 273–290.

Warr, N., Powles-Glover, N., Chappell, A., Robson, J., Norris, D., & Arkell, R. M. (2008). Zic2-associated holoprosencephaly is caused by a transient defect in the organizer region during gastrulation. *Human Molecular Genetics, 17*(19), 2986–2996. https://doi.org/10.1093/hmg/ddn197 PMID: 18617531.

Wilson, S. W., & Houart, C. (2004). Early steps in the development of the forebrain. *Developmental Cell, 6*(2), 167–181. https://doi.org/10.1016/s1534-5807(04)00027-9 PMID: 14960272; PMCID: PMC2789258.

Winklbauer, R. (1990). Mesodermal cell migration during Xenopus gastrulation. *Developmental Biology, 142*, 155–168.

Wolff, C., Roy, S., & Ingham, P. W. (2003). Multiple muscle cell identities induced by distinct levels and timing of hedgehog activity in the zebrafish embryo. *Current Biology, 13*, 1169–1181.

Yamada, T. (1994). Caudalization by the amphibian organizer: Brachyury, convergent extension and retinoic acid. *Development (Cambridge, England), 120*, 3051–3062.

Yamada, T., Placzek, M., Tanaka, H., Dodd, J., & Jessell, T. M. (1991). Control of cell pattern in the developing nervous system: Polarizing activity of the floor plate and notochord. *Cell, 64*(3), 635–647. https://doi.org/10.1016/0092-8674(91)90247-v PMID: 1991324.

Yoshihi, K., Kato, K., Iida, H., Teramoto, M., Kawamura, A., Watanabe, Y., et al. (2022). Live imaging of avian epiblast and anterior mesendoderm grafting reveals the complexity of cell dynamics during early brain development. *Development (Cambridge, England), 149*(6), dev199999. https://doi.org/10.1242/dev.199999.

Zhang, X., Cheong, S. M., Amado, N. G., Reis, A. H., MacDonald, B. T., Zebisch, M., ... He, X. (2015). Notum is required for neural and head induction via Wnt deacylation, oxidation, and inactivation. *Developmental Cell, 32*, 719–730.

Zhou, X., & Anderson, K. V. (2010). Development of head organizer of the mouse embryo depends on a high level of mitochondrial metabolism. *Developmental Biology, 344*(1), 185–195. https://doi.org/10.1016/j.ydbio.2010.04.031.

Zoltewicz, J. S., & Gerhart, J. C. (1997). The Spemann organizer of Xenopus is patterned along its anteroposterior axis at the earliest gastrula stage. *Developmental Biology, 192*(2), 482–491. https://doi.org/10.1006/dbio.1997.8774.

CHAPTER FIVE

Transport and gradient formation of Wnt and Fgf in the early zebrafish gastrula

Emma J. Cooper and Steffen Scholpp*
Living Systems Institute, Faculty of Health and Life Sciences, University of Exeter, Exeter, United Kingdom
*Corresponding author. e-mail address: s.scholpp@exeter.ac.uk

Contents

1. Introduction	126
2. The discovery of the Spemann-Mangold Organiser	127
3. Morphogen signalling from the organiser	128
4. Comparing the Wnt and Fgf signalling pathways on a molecular level	129
5. Contrasting the role of Wnt and Fgf signalling within embryonic patterning	130
5.1 Neural induction	130
6. Anteroposterior axis formation	131
7. Neural AP axis patterning	132
8. Mesodermal and endodermal fate	133
9. Juxtaposing the transport mechanisms for Wnt and Fgf	134
10. Post-translational modification	134
11. Carrier proteins	135
12. Restricted diffusion by heparan sulphate proteoglycans (HSPGs)	135
13. Extracellular vesicles	138
14. Cytonemes facilitate paracrine morphogen signalling	139
15. Comparing Wnt and Fgf signalling gradient formation	141
15.1 Gradients through control and clearance	141
16. Controlled morphogen transport shapes the gradient	142
17. Signalling modulators in the target cells	143
18. Final remarks on the importance of transport modes in morphogen gradient formation	144
Acknowledgement	146
References	146

Abstract

Within embryonic development, the occurrence of gastrulation is critical in the formation of multiple germ layers with many differentiative abilities. These cells are instructed through exposure to signalling molecules called morphogens. The secretion of morphogens from a source tissue creates a concentration gradient that allows distinct pattern formation in the receiving tissue. This review focuses on the morphogens

Wnt and Fgf in zebrafish development. Wnt has been shown to have critical roles throughout gastrulation, including in anteroposterior patterning and neural posterisation. Fgf is also a vital signal, contributing to involution and mesodermal specification. Both morphogens have also been found to work in finely balanced synergy for processes such as neural induction. Thus, the signalling range of Wnts and Fgfs must be strictly controlled to target the correct target cells. Fgf and Wnts signal to local cells as well as to cells in the distance in a highly regulated way, requiring specific dissemination mechanisms that allow efficient and precise signalling over short and long distances. Multiple transportation mechanisms have been discovered to aid in producing a stable morphogen gradient, including short-range diffusion, filopodia-like extensions called cytonemes and extracellular vesicles, mainly exosomes. These mechanisms are specific to the morphogen that they transport and the intended signalling range. This review article discusses how spreading mechanisms in these two morphogenetic systems differ and the consequences on paracrine signalling, hence tissue patterning.

1. Introduction

Throughout the history of life sciences, there has been a fascination with embryonic development and the process of the transformation of a single cell into a multicellular differentiated organism. To allow this, the simple arrangement of cells – the so-called blastula – is transformed into a complex embryo, including three germ layers of the gastrula. The germ layers, named the ectoderm, mesoderm, and endoderm, undergo induced differentiation and altered morphology to form all cell types. The ectoderm contributes to the nervous system and the skin, the mesoderm is essential for muscle formation, and the endoderm allows the formation of many inner organs. The development of these germ layers is highly conserved throughout species. However, the mechanism behind this formation varies significantly between species. A crucial tool orchestrating the development of the germ layers of the gastrula is chemical signalling through form-giving substances, so-called morphogens. Cells use these morphogens to induce the germ layers and assign positional information to the individual cells. Thus, these signalling factors can control the shape and form (Greek: *morphe*) of an organism. Surprisingly, these signalling proteins are made by a relatively small group of cells, forming a gradient in the neighbouring tissue to organise their development in a concentration-dependent manner. Vital for their mode of action is dissemination from the producing cells to the receiving cells. This review compares two crucial morphogenetic signalling systems: the Wnt and the Fgf signalling cascades. It contrasts how their signalling components – although produced in the same source cells – spread through an early embryonic tissue, here the zebrafish gastrula. An advanced understanding of the distribution of morphogens in gastrulation is crucial to fully comprehending how a signalling gradient is established and, thus, the first and essential steps in forming complex multicellular organisms.

2. The discovery of the Spemann-Mangold Organiser

The finding by Hilde Mangold and Hans Spemann significantly enhanced our understanding of early embryogenesis. About 100 years ago, Mangold conducted decisive experiments on amphibian embryos, leading to her doctorate and later to winning the Nobel Prize for Spemann. Their research hypothesised that, on the one hand, a group of cells, the so-called central (or primary) organiser, orchestrates vital embryonic processes such as the induction, arrangement, and differentiation of the surrounding tissue into the three germ layers (Spemann & Mangold, 1924). On the other hand, they postulated that the neighbouring tissues could be characterised by their "ability to react (reaktionsfähig)" upon the signals from the organiser. Using two newt species, they explored this interplay between the sender and receiver tissue.

The dorsal blastopore lip (formed through invagination) of a *Triturus taeniatus* embryo was removed and transplanted under the ectoderm within the ventral region of a source of *Triturus cristatus*, a similar but unpigmented species. The transplanted dorsal lip mainly differentiated into a notochord, whereas the overlying host ectoderm was induced to differentiate into the neural plate. Somites were found to consist of both unpigmented tissue from the host and pigmented donor tissue. After allowing gastrulation to continue, a second embryo was formed, conjoined to the larger donor embryo. After this discovery, the dorsal lip and its derivatives have been penned as the 'Spemann-Mangold organiser'. These findings by Hilde Mangold and Hans Spemann significantly enhanced our understanding of early embryogenesis. Their initial concept has been expanded and refined by Conrad Waddington by the term "competence", which has been defined as the ability of a cell or tissue to respond to an inducing signal by changing its cellular fate (Waddington, 1940). Today, competence does not only include the ability to "see" the morphogen (presence of the appropriate receptors) but also the presence of the intracellular signalling cascade, including key transcription factors and the state of the chromatin controlling the transcriptional profile.

3. Morphogen signalling from the organiser

In the 1950s, Alan Turing hypothesised that the patterns we observe during embryonic development arise in response to a spatial pre-pattern in biochemicals, which he termed morphogens (Turing et al., 1952). These morphogens are produced by a signalling source and are vital for cell fate specification and embryo organisation. A decade later, Wolpert built the French flag model on these previous findings and provided a widely accepted tissue patterning theory (Wolpert, 1969). It proposes that patterning occurs through 'positional and spatial information' and that tissue differentiates according to the informative signals it receives. In detail, a signalling molecule can only be defined as a morphogen when the receiving cells respond to the signal with at least two different responses according to the morphogen concentration they are exposed to. As the gradient of information changes throughout the embryo, the tissue becomes differentiated in a parallel fashion. It was soon found that specific morphogens provided an answer to the 'informative signals' that Turing and Wolpert proposed. The discovery of the Spemann organiser furthered this theory by providing a source from which morphogens are secreted and gradients can be formed.

In subsequent works, it has become clear that the central (or primary) organiser is vital in allowing gastrulation to occur. To allow the formation and function of the organiser, several signalling systems are required in a precise spatiotemporal controlled activity throughout development. The initial signals belong to the pathways of Nodal, Bone morphogenetic protein (BMP), Wnt, and Fibroblast growth factors (Fgf). Nodal specifies endoderm and mesoderm and patterns the germ layers, and BMP defines ventral fates and patterns in the dorsal/ventral axis. In the embryo, Nodal signals and BMP antagonists, secreted from the Spemann–Mangold organiser, induce anterior neural tissue, dorsal mesoderm, somites, and notochord (Zhou et al., 1993; reviewed in DeRobertis & Kuroda, 2004). In parallel, Wnts and Fgfs produce a patterning gradient in the neural plate through interactions with antagonistic molecules, controlling gene transcription and, ultimately, cell fate (Briscoe & Small, 2015; PLACE-HOLDER - > CROSS-REFERENCES to reviews from C. Stern and J. Slack in this issue). This review will focus on these two growth factor families, the Wnt and Fgf signals and their involvement in orchestrating the controlled differentiation required within embryonic development in zebrafish.

4. Comparing the Wnt and Fgf signalling pathways on a molecular level

Since the first description of Wnt protein family members, Int-1 in mice (Nusse & Varmus, 1982), and Wingless (Wg) in Drosophila (Nüsslein-Volhard & Wieschaus, 1980), the Wnt signalling cascade has undergone intensive research. In humans, 19 Wnt proteins interact with 10 Frizzled family receptors and several co-receptors (Niehrs, 2012). The binding of Wnt ligands to these receptors produces vital paracrine signalling pathways essential throughout life from embryogenesis to adult tissue homeostasis. Two major signalling pathways can be activated by binding the ligand: the canonical and the non-canonical. The most widely understood pathway is the canonical, which relies on stabilising the intracellular signal transducer β-catenin (Nusse & Clevers, 2017). In the absence of Wnt, β-catenin is phosphorylated by GSK3 within the β-catenin destruction complex (comprising the kinases GSK3 and CK1 and the scaffolding proteins APC and Axin). This phosphorylation tags β-catenin for ubiquitination, followed by proteasomal destruction (Fig. 1A).

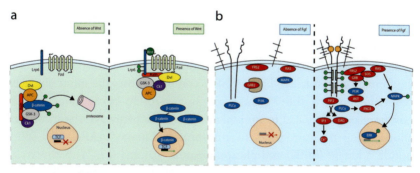

Fig. 1 The signalling pathways. (A) The canonical Wnt pathway - Upon binding a Wnt ligand with the Frizzled receptor (Fzd) and the co-receptor (Lrp6), the components of the destruction complex, Dishevelled (Dvl) and Axin, together with the kinases Gsk3β, CK1 and the scaffolding protein APC, are recruited to the membrane. The conformational changes of Axin inhibit the phosphorylation (green P) of β-catenin and allow its cytoplasmic accumulation. β-catenin can then translocate into the nucleus and regulates target gene transcription through interaction with TCF/LEF transcription factors. (B) For activation, FgfRs are bound by two Fgf ligands and two HSPG molecules, which lead to the transduction of intracellular pathways. The three major pathways that can be activated through the FgfR receptor are MAPK/ERK, controlling cell proliferation and differentiation; the PI3K pathway, regulates cell survival via AKT; and the PLCγ pathway, which affects cell morphology and migration via the release of Calcium.

In the presence of a Wnt, the ligand binds to both the bona fide receptors of the Frizzled (Fzd) family and specific co-receptors, such as Lrp5/6, forming a cluster of trimeric complexes that recruits components of the β-catenin destruction complex such as the scaffolding proteins, Dishevelled and Axin to the membrane. This recruitment inhibits the function of the destruction complex, and hence prevents β-catenin phosphorylation. As a result, β-catenin, no longer ubiquitinated, is stabilised, allowing its accumulation and translocation into the nucleus. Upon entering the nucleus, β-catenin interacts with the Lef/Tcf transcription factors, regulating downstream gene expression. For completion, the non-canonical Wnt pathways rely on the remodelling of the cytoskeleton to influence cell polarity and migration (Wnt/PCP pathway) and changes in Calcium signalling (Wnt/Ca^{2+} pathway; Yang & Mlodzik, 2015).

In addition to Wnt signalling, Fgfs are also crucial morphogens within embryonic development and are highly conserved throughout species (Teven et al., 2014; Ornitz & Itoh, 2015). They play a key role in cell proliferation, differentiation, and patterning. Within mammals, 18 proteins within the Fgf family act as signalling ligands to 4 Fgf receptors (FgfRs) to transduce multiple signalling pathways. For Fgf signalling to occur, a complex must be formed on the plasma membrane (Fig. 1B). For example, the Fgf2-FgfR1 complex consists of two Fgf ligands, two receptors and two heparin sulphate proteoglycan (HSPG) cofactors (Mohammadi et al., 2005). The formation of this Fgf-FgfR-HSPG complex allows the intracellular co-phosphorylation of the FgfRs through a tyrosine-kinase domain. This phosphorylation can lead to multiple transduction pathways, the most well-defined being the mitogen activating-protein kinase (MAPK), phosphoinositide 3 kinase (PI3K) and phospholipase C gamma (PLCγ) pathways.

5. Contrasting the role of Wnt and Fgf signalling within embryonic patterning

5.1 Neural induction

In early gastrulation, ectoderm must be instructed to differentiate into neural or epidermal ectoderm (PLACEHOLDER -> CROSS-REFERENCES reviews from C. Stern and J. Slack in this issue). Primary research within Xenopus embryos has provided evidence that the expression of the morphogen BMP causes pluripotent ectodermal cells to adopt an epidermal fate. Antagonists of BMP such as Noggin and Chordin are released from the organiser to induce a neural fate within ectodermal cells (Lamb et al., 1993; Sasai et al., 1995). Deleting these BMP antagonists leads to the fatal loss of dorsal structures and the Xenopus neural plate (Khokha et al., 2005). Although this model explains neural induction within Xenopus, within the chick embryo Chordin-mediated inhibition of BMP was insufficient to form neural ectoderm (Streit et al., 1998), suggesting that the induction of neuroectoderm may require additional input. Further studies have provided evidence that Fgf signalling, in addition to BMP antagonists, is required for successful neural induction (Kengaku & Okamoto, 1993; Streit et al., 2000; Delaune et al., 2005). However, the mechanism by which Fgf signalling results in neural induction has yet to be fully understood, and several theories have been proposed. One potential mechanism is BMP inhibition through the Fgf-MAPK-Smad1/2 pathway (Pera et al., 2003). Another proposed mechanism of BMP inhibition is through the Fgf-MAPK-Ets pathway. The transcription of Ets proteins through the transduction of the MAPK pathway is thought to upregulate genes that determine a neural fate in ectodermal cells (Hongo & Okamoto, 2020, 2022; Wasylyk et al., 1998). It is now widely accepted that both BMP and Fgf signalling cooperate to produce correct neural induction (Kudoh et al., 2004).

More recently, Wnt signalling has also been found to be involved in deciding ectoderm cell fate. However, conflicting research has failed to elucidate the mechanism (Brafman & Willert, 2017). One proposed method is that Wnt signalling is an antagonist towards BMP; therefore, its posterior secretion inhibits neural specification (Baker et al., 1999). However, studies have also shown that Wnt can indirectly stimulate BMP secretion through interaction with Fgf and cause epidermal specification (Wilson et al., 2001).

6. Anteroposterior axis formation

After formation, the organiser secretes multiple factors that mediate germ layer organisation and the anteroposterior axis (AP axis). Wnt signalling is the primary pathway that instructs the formation of this axis (Kiecker & Niehrs, 2001; Nordström et al., 2002). During gastrulation, the organiser (comprising axial mesodermal cells) migrates and elongates anteriorly towards the animal pole (Yanagi et al., 2015). It then secretes multiple morphogen antagonists, including the Wnt antagonists Dkk1, Cerberus, and sFRPs (Glinka et al., 1998; Piccolo et al., 1999; Houart et al., 2001). These factors inhibit β-catenin stabilisation either through direct binding of the Wnt ligand or through binding to the Lrp5/6 co-receptor, inhibiting their functionality (Yamaguchi, 2001). Within the posterior region, high levels of Wnt are expressed with evidence of Wnt3, 3a, 5a, 5b, 8a and 11 being present throughout species (Ding et al., 2017; Andre et al., 2015; Lekven et al., 2001; Takada et al., 1994). The interaction of posterior Wnt signalling with anterior antagonism creates a gradient of active Wnt ligands. Research has shown that upregulation of anterior Wnt causes abnormal head formation, and Wnt inhibition upregulation leads to a secondary axis as well as stunted trunk and tail growth (Bouwmeester et al., 1996; Christian & Moon, 1993).

7. Neural AP axis patterning

To induce neuroectoderm patterning along the AP axis, the Wnt signalling gradient causes the posteriorisation (or caudalisation) of many neural cells that are fated to become the midbrain and hindbrain. It has been shown that inhibition of several Wnts (specifically, Wnt8 and Wnt3a) led to overexpansion of anterior neuroectoderm within the forebrain and a lack of posterior growth (Shimizu et al., 2005). Wnt signalling has also been found to be crucial in the development of the forebrain. In order for the posterior diencephalon to develop, Wnt signalling is necessary. At the same time, the anterior telencephalon cell differentiation required Wnt antagonism (Heisenberg et al., 2001; Houart et al., 2002). The telencephalon is also dorsalised through Wnt signalling by Wnt ligands (Hébert & Fishell, 2008). Parr et al. (1993) demonstrated that Wnt1, 3, 3a, 4 and 7b were expressed dorsally, whereas Wnt5a and Wnt7a were expressed ventrally.

Complementary to the Wnts, Fgf-soaked beads inserted into embryos led to the formation of ectopic posterior neural structures, specifically posterior hindbrain and spinal cord, and the deletion of anterior structures within the chick and Xenopus (Ribisi et al., 2000; Alvarez et al., 1998). Modification of FgfRs, causing them to be non-functional, led to a shortened axis and significantly impaired posterior structures (Scholpp et al., 2004; Ota et al., 2009). Previously induced anterior neuronal cells have also been shown to get posteriorized after exposure to Fgf signalling (Lamb & Harland, 1995). In addition, multiple genes that are considered to encourage a posterior fate are found to be upregulated after overexpression of Fgf signals, such as *xcad3*, *hoxB9* and *hoxA7* (Lamb & Harland, 1995; Pownall et al., 1996). This demonstrates the importance of Fgf signalling in regulating HOX genes and, therefore, axis patterning and body segmentation.

In addition, and similar to Wnt, Fgf has also been found to have a role in brain patterning. First evidenced in 1994 by Heikinheimo et al. (1994), Fgf8 is present within the midbrain and hindbrain (MHB) boundary. This signalling centre, known as the isthmus centre, is thought to be responsible for tissue specificity and patterning within the brain. Insertion of Fgf8-soaked beads into the diencephalon replicated transplanted isthmus centre activity (Crossley et al., 1996). Furthermore, misexpression experiments deduced that Fgf8 was found to have a strong transforming power by upregulating several genes, including *Gbx2, Pax2 and Irx2,* whilst repressing *Otx2* expression. This led tissue with tectum and caudal diencephalon fate to transform to have cerebellum characteristics (Reifers et al., 1998; Sato et al., 2001). These findings suggest that Fgf signalling is a critical event that controls the fate of neural tissue within the midbrain and hindbrain area.

8. Mesodermal and endodermal fate

Signalling events induced by the Fgf ligand are essential throughout embryonic development. One initial event that relies on Fgf signalling is the formation of germ layers through the differentiation of epiblast cells via epithelial-mesenchymal transition (EMT). This process allows tightly held, rigid epithelial cells to adopt mesenchymal characteristics allowing their migration and invasion after EMT (Thiery et al., 2009; Kalluri & Weinberg, 2009). During gastrulation, epiblast cells ingress through the primitive streak using EMT and differentiate to a mesodermal or endodermal fate.

The role of Fgf signalling within embryonic development was initially explored through loss-of-function studies. Targeted disruption of Fgfs and their receptors within mouse and Xenopus embryos led to growth impairment and lethality through the reduced and aberrant formation of mesodermal tissue (Slack et al., 1996; Yamaguchi et al., 1994; Sun et al., 1999). Further research using chimeric analysis demonstrated that FgfR$^{-/-}$ cells accumulate within the primitive streak, failing to induce mesodermal and ectodermal fates (MacNicol et al., 1993; Umbhauer et al., 1995; LaBonne et al., 1995). Sun et al. (1999) observed that Fgf8 Is required after EMT, for cellular migration away from the primitive streak and is also required for the induction of Fgf4 expression. Within mice, FgfR1 expression is required for the expression of *Snail*, a transcription factor necessary for E-cadherin downregulation (Ciruna & Rossant, 2001). Taken together, this evidence suggests that expression of Fgf and its receptors are needed to signal intercellular transduction pathways such as Ras/MAPK, which subsequently produce transcription factors required for EMT, allowing ingression and ultimately leading to a mesodermal or endodermal fate.

9. Juxtaposing the transport mechanisms for Wnt and Fgf

The evidence detailed above demonstrates that both the Wnt and Fgf morphogens are integral in tissue patterning during gastrulation through their dissemination. However, an important question yet to be answered is how these morphogens can create this gradient throughout the tissue. Numerous studies have produced several theories that propose an answer to this question, which will be summarised in this section.

10. Post-translational modification

Wnts and Fgfs are paracrine signalling proteins, and their post-translational modifications are essential for folding and stability but also influence their paracrine signalling activity. Although Wnt signalling is crucial throughout development, it challenges the classical term of a 'morphogen' as its post-translation modifications make it hydrophobic, thus, diminishing its ability to diffuse passively in the extracellular space. Therefore, numerous methods are thought to be relied upon for Wnt gradient formation. Upon translation of a Wnt protein, it undergoes posttranslational palmitoleation within the ER by a membrane-bound O-acetyltransferase porcupine (PORCN; Alvarez-Rodrigo et al., 2023). This modification is required to allow the binding of Wnt to its receptor Frizzled (Janda et al., 2012). However, as this O-acylation causes Wnts to become highly hydrophobic, they require mediated transport from the ER (Proffitt & Virshup, 2012; Zhai et al., 2004). This trafficking is accomplished by a dedicated Wnt transporter, the integral membrane protein Evi/Wls. The binding of palmitoleated Wnt to Wls allows their transport to the plasma membrane via the Golgi, where further modification of glycosylation occurs (Nygaard et al., 2021; Mehta et al., 2021).

There is also evidence that Fgfs can be post-translationally modified. For example, glycosylation is important for the cell-surface transport of Fgfs

(Guan et al., 1985) and substitution of the Fgf3 amino-terminus with that of Fgf5 led to the efficient secretion of the hybrid protein in cell culture. Furthermore, in Drosophila, there is evidence of post-translational modification of the Fgf homolog Branchless (Bnl) by multiple proteolytic cleavages and the addition of a glycosylphosphatidylinositol (GPI) moiety that can anchor Fgf/Bnl on the producing cell surface (Du et al., 2022).

11. Carrier proteins

One of the major mechanisms in which morphogens, like Wnt and Fgfs, can traverse through the embryo is through binding with a carrier protein. In the case of Wnt proteins, these carrier proteins function by 'shielding' the hydrophobic moiety of a Wnt ligand during extracellular transport (Fig. 2A). One example of a proposed carrier protein is the sFRP family consisting of proteins such as Frzb and Crescent. These proteins were initially found to be secreted from the organiser, acting antagonistically towards Wnt and influencing anterior structure (Leyns et al., 1997; Wang et al., 1997). However, recent research has suggested that the sFRP family behaves differently depending on its local concentration and on the distance from the signalling source. For example, Crescent has been shown to diffuse posteriorly and interact with stagnant posterior Wnts to form a Wnt-sFRP complex, allowing Wnt transportation towards the anterior region (Mii & Taira, 2009) and at lower concentrations, sFRPs can facilitate signalling (Uren et al., 2000; Xavier et al., 2014). Furthermore, recent research has uncovered that the serum glycoprotein Afamin forms a 1:1 complex with lipidated Wnts and aids in the solubility of Wnts in the aqueous extracellular space (Mihara et al., 2016). In addition, the secreted wingless-interacting molecule (Swim), a member of the Lipocalin family, was found to bind Wg with a nanomolar affinity, maintaining its solubility and response to Fzd ligands (Mulligan et al., 2012).

12. Restricted diffusion by heparan sulphate proteoglycans (HSPGs)

The heparan sulphate proteoglycans (HSPGs) are an additional and important class of extracellular interactors (Fig. 2). HSPGs are glycoproteins with the common characteristic of containing one or more covalently

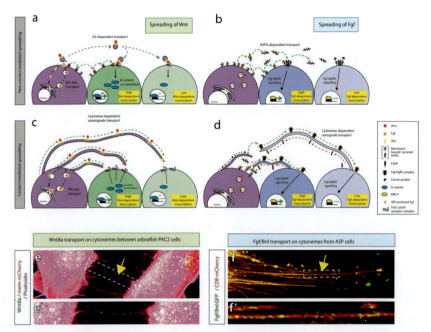

Fig. 2 Intercellular transport mechanisms of Wnt and Fgf ligands. (A) Glypican-assistant and exosome-mediated transport of Wnt. After Wls-mediated trafficking of Wnt to the plasma membrane, Wnt can bind extracellularly to Glypicans, facilitating its short-range transport (green arrows). After the release from multi-vesicular bodies, Wnt can also travel on exosomes (blue arrows) to the receiving cells. (B) After the release, Fgfs can be stored on HSPGS, i.e. Perlecans, in the extracellular space, and Fgf signals can use HSPGs for short-range transport (green arrows). (C) Wnt proteins can be transported on cytonemes from the source cell to the receiving cell (green arrow). (D) Cytonemes can also mobilise the FgfR to pick up the Fgf ligand at a distance (red arrows) and transport it back to the receptor-producing cell (green arrows). Similarly, the Fgf ligand can be transported along cytonemes (black arrows). (E) Antibody staining against Wnt8a (green) in mem-mCherry (red) transfected zebrafish fibroblasts and counterstained with Phalloidin (blue). Between PAC2 cells, the Wnt8a protein is visible on cytonemes and at cytoneme contact sites (yellow arrows). (E′) zoomed-in area indicated by the ROI box in (E). (F) Example for Fgf transport along cytonemes: Enhanced-gain super-resolution images, showing a sub-resolution distribution of Fgf/Bnl: GFP molecules in live CD8:Cherry-marked air sac primordium (ASP) cells and ASP cytonemes in Drosophila; genotype: btl-Gal4, UAS-CD8:Cherry/+; bnl:gfp^{endo}; (F′) zoomed-in area indicated by the ROI box in (F). *Courtesy S. Roy; University of Maryland, USA; published in Du, L., Sohr, A., Yan, G., & Roy, S. (2018). Feedback regulation of cytoneme-mediated transport shapes a tissue-specific FGF morphogen gradient. Elife, 7, e38137.*

attached heparan sulphate (HS) chains. This well-studied family is known to bind to an abundance of ligands by the HS side chains, including multiple morphogens. Therefore, there is an enormous interest in the role of their interactions with Fgf ligands (Fig. 2B). HSPGs consist of a core protein with varying numbers of attached glycosaminoglycans (GAGs), the HS chains. Three major classes of HSPGs have been identified with varying compositions and cellular positions. Glypicans and syndecans span the membrane, with perlecans being secreted into the ECM. Original research into the relationship between Fgf and HSPGs demonstrated that Fgfs have a high affinity to heparin, leading to increased stabilisation and signalling (Ornitz, 2000). For example, in the presence of HSPGs, the degradation rate of Fgf is reduced, and therefore the radius of target cells is increased (Gospodarowicz & Cheng, 1986; Flaumenhaft et al., 1990). There is also strong evidence that Fgf8 uses HSPGs to spread in the zebrafish embryo. Studies using single-molecule imaging found for Fgf8 in the zebrafish gastrula, a slow-moving population ($4 +/- 3\,\mu m^2/s$) and a fast-moving population ($53 +/- 8\,\mu m^2/s$), suggesting that single molecules of Fgf8 can be detected in the extracellular space as and they move by free Brownian diffusion (Yu et al., 2009). Changing the HSPGs composition or the degree of uptake of Fgf8 into its target cells alters the shape of the Fgf8 gradient (Scholpp & Brand, 2004; Nowak et al., 2011). Similarly, a mutation within the *Fgf9* gene of mice prevents the protein from homodimerising, decreasing its affinity to the heparin chains of HSPGs. The diffusion rate of Fgf9 is then less spatially controlled, which allows its ectopic signalling throughout the embryo, leading to developmental abnormalities (Harada et al., 2009). After Fgf-HSPG binding, an HSPG cleaving process in Xenopus embryos is thought to aid in ligand release and signalling gradient (Hou et al., 2007). Interestingly, modulation of HSPGs may also affect gradient formation differently: HSPGs can affect cytoneme formation in Drosophila that communicate Fgf/Bnl and Wnt/Wg (Huang and Kornberg et al., 2015).

In addition to aiding in controlling Fgf spreading, the heparin sulphate (HS) chains of HSPGs are also thought to act as co-receptors forming an HS-Fgf-HSPG ternary complex. Pioneering studies have determined that preventing low-affinity HSPG binding with Fgf reduces the high-affinity binding with the FgfRs, hence a decrease in signalling (Fig. 2B; Yayon et al., 1991; Rapraeger et al., 1991).

HSPGs have also been found to be necessary for Wnt transportation (Fig. 2A). In Xenopus, HSPGs are locally clustered on cell surfaces.

Depending on the side chains, HSPGs can accumulate Wnt8 and alter distribution and signalling (Mii et al., 2017). In Drosophila, two glypican proteins, Dally and Dlp, are involved in Wnt signalling. Drosophila mutant clones with null alleles for Dally and Dlp result in a decrease in extracellular Wg and a loss of efficient Wnt/Wg gradient formation (Han et al., 2005; McGough et al., 2020). Specifically, Dlp captures Wnt/Wg to facilitate paracrine signalling (Franch-Marro et al., 2005). The importance of the GAG chain within HSPGs has been shown multiple times using genetic mutants of UDP glucose dehydrogenase (or invertebrate homologue), an enzyme required for the biosynthesis of GAGs. *Sugarless (Sgl)* and *Kiwi* are examples of these genes that have been ablated. In both Drosophila mutants *for Sgl* and *Kiwi*, Wnt/Wg abundance is reduced, evidencing a reliance on GAG during processing. These studies suggest that HSPGs aid in Wnt signalling by maintaining its localisation and concentration at the cell surface for interaction with the Fzd receptor (Yan & Lin, 2009; Lin, 2004). Recent evidence suggests that glypicans bind directly to Wnt/Wg, as the palmitoleate on Wg can interact with a groove-like domain on the Drosophila glypican Dlp comparable to the one found in the cysteine-rich domain of the Wnt receptors (McGough et al., 2020). By interaction of morphogens with ECM components, the signalling components may be stored in the ECM and can travel between neighbouring cells.

13. Extracellular vesicles

The loading of morphogens onto extracellular vesicles (EVs) has been a popular theory for efficient transport. An interaction between Wnt and EVs is specifically appealing as they allow the concealment of the hydrophobic moiety (Fig. 2A). The concept of extracellular vesicle transport for Wg was initially discovered within Drosophila (Greco et al., 2001). An association between basolateral-derived lipoproteins and Wg membrane accumulation was found. This paper also proposed interaction with HSPGs and lipoproteins, suggesting that these mechanisms of Wnt transport may rely on interactions between each other.

Within Drosophila, Wg is found to colocalise with lipophorin (invertebrate homologue of lipoprotein particles), and Wg levels were reduced in larvae with downregulated lipoprotein levels (Panáková et al., 2005). In vertebrates, Wnt5a can be mobilised on lipoproteins in the cerebrospinal canal in the hindbrain in mice (Kaiser et al., 2019). However, conflicting evidence has found that overexpression of membranous Wg does not affect membrane lipophorin levels, suggesting that Wg does not rely on lipid-based extracellular vesicles for transport (McGough et al., 2020).

Exosomal transport is now a well-recognised mechanism for Wnt transport on EVs (Fig. 2A). In the neuromuscular junction in Drosophila, Wnt/Wg, together with Wls, is transported on exosomes (Korkut et al., 2009). Biochemical fractional analysis has found that Wg and Wnt3a co-segregate with exosomal markers such as Tsg101 and are active in the receiving cells (Gross et al., 2012). Since these studies, exosomes have been shown to contribute to Wnt/Wg spreading in many contexts, including in diseases, through activating aberrant Wnt signalling (Luga et al., 2012; Działo et al., 2019).

There is also some evidence that EVs can distribute Fgf. For example, a recent report indicates that Fgf2 can be bound on the surface of EVs to alter signalling in the receiving cells (Petit et al., 2022). However, more evidence is needed to propose EVs as an essential propagation mechanism for Fgfs as they are for Wnts.

In summary, EVs are one way to mobilise morphogens in a tissue. However, there are still remaining questions. For example, how this mechanism can target specific cells is still being determined. Furthermore, evidence must be provided to explain how the ligand is placed on the surface of an EV to be able to interact with its receptor on the plasma membrane of the receiving cell.

14. Cytonemes facilitate paracrine morphogen signalling

A more novel proposal of the mechanism of morphogen transport is through their presentation on signalling protrusions, also known as cytonemes (Fig. 2). They were first discovered by Ramírez-Weber and Kornberg (1999) within the Drosophila wing imaginal disc. These can be actin-rich filopodia-like structures or microtubule-containing nanotubes, which polarise towards morphogen-producing regions, extending up to 100 μm. In addition, cytonemes can also carry the ligand towards the receiving cell (Kornberg & Roy, 2014; Zhang & Scholpp, 2019).

One of the best-characterised examples of cytoneme-mediated transport is spreading Wnt proteins. In zebrafish embryogenesis, Wnt and its related proteins have been found on cytonemes and initiate Wnt signalling on neighbouring cells through direct physical contact with the cytoneme tip (Fig. 2C; Stanganello et al., 2015; Stanganello & Scholpp, 2016). The formation of these cytonemes is controlled through Wnt molecules, which act in an autocrine fashion through interaction with Ror2 which, in turn, activates the Wnt/PCP signalling pathway. This leads to regulating multiple small GTPases, such as Rho and Cdc42, thus promoting filopodia formation through actin polymerisation and organisation at the membrane. Wnt ligands can then be loaded onto cytonemes and bind to Fzd receptors on neighbouring cells, initiating the β-catenin-dependent Wnt pathway (Mattes et al., 2018). Wnt cytonemes can reach a length of up to 100 μm in the zebrafish embryo or in gastric tumouroids (Stanganello et al., 2015; Rogers et al., 2023). In addition to Wnt on cytonemes, there is evidence that other Wnt signalling components proteins can be found on cytonemes. For example, Vangl2 and Ror2 can also be loaded onto cytonemes, in addition to Wnt, and activate JNK signalling, which in turn increases cytoneme length (Brunt et al., 2021). The action of these proteins is now considered a key process for the spatiotemporal control of Wnt signalling within embryonic development. Complementary to Wnt signalling components, further important proteins regulating cytoneme-related Wnt signalling are membrane scaffolding proteins like Flotillin2 (Flot2). Flot2 is a membrane-bound scaffolding protein that has been found to promote filopodia formation. Flot2 also marks specific lipid domains in the external leaflet of the plasma membrane – also known as lipid rafts – which are enriched in cholesterol, glycosphingolipids, and glycosyl-phosphatidylinositol GPI-anchored proteins. In gastric cancer cells, Flot2 promotes the dissemination of Wnt3a. It thus facilitates the maintenance and proliferation of the cancer stem cell (Routledge et al., 2022), suggesting that the correct lipid composition of the membrane is a prerequisite for cytoneme formation.

The concept of the extension of ligand-bearing cytonemes to a receiving cell has been expanded by studies in invertebrates. Within Drosophila, Wg has not been found on these structures, but instead, studies have shown the presence of Fzd receptor-loaded cytonemes (Huang & Kornberg, 2015). This suggests that the receiving cell produces these extensions to find a membrane-bound Wg ligand (Fig. 2C).

Recent data suggests that cytonemes can not only tranport Wnt ligands but also Wnt receptors. The discovery that Wnt5b and Ror2 move together on cytonemes to the receiving cell is astonishing and unveiled a remarkable synergy between transport and signalling (Zhang et al., 2023). The ability to initiate signalling in the receiving cell - independent of receptor expression – may challenge the idea of cellular competence, which is based on the assumption that signalling depends foremost on the availability of the correct receptors.

In Drosophila, strong evidence suggests that the Fgf can be transported along cytoneme extensions (Roy et al., 2011, 2014; Du et al., 2018, 2022; Patel et al., 2022). Work on Drosophila air-sac primordium showed that Fgf-exchanging cytonemes are extended by both source and recipient cells and exchange signals at their point of contact (Fig. 2D). Fgf/Bnl activate the receptors on the cytonemes. The observation that cytonemes from ASP cells forms tight contact sites with the Fgf/Bnl releasing cells led to the intriguing comparison of cytoneme contact sites with membrane synapses (Kornberg & Roy, 2014; Huang et al., 2019). Recently, there is evidence that Fgf/Bnl can be membrane-anchored by adding a glycosylphosphatidylinositol moiety (Du et al., 2022), which both would reduce free Fgf secretion and promote its cytoneme-dependent transport. This allows ligand-producing cells to use cytonemes to establish contact with FgfR-expressing cells or cytonemes selectively via Fgf-FgfR binding, similar to the interactions of cell-adhesion molecules. Fgf-FgfR binding is followed by target-specific Fgf exchange at the contact sites. However, the mechanism of Fgf release from the GPI anchor is unknown. Fgf reception and signalling activation induce dose-dependent variable responses in the recipient cells, influencing cytoneme formation, polarity, and contacts. In addition, work in Drosophila adult muscle progenitor niche identified two other Fgfs in Drosophila, Pyramus/Pyr and Thisbe/Ths, and their receptor Heartless (Htl) localise on cytonemes. Cytoneme-mediated Pyr/Ths signalling maintains the niche by a positive feedback loop. An increase in Bnl/Fgf signalling in the recipient cells leads to an increase in cytoneme formation, which, in turn, polarises FgfR/Htl-containing recipient cytonemes to selectively polarise toward the Fgf/Pyr/Ths expressing niches adhere to the niche to maintain stemness (Patel et al., 2022). These findings provide insights into how crucial cytoneme-mediated Fgf signal can generate an Fgf gradient in the tissue in invertebrates. It will be interesting to see if Fgf and their receptors are similarly present on cytonemes in vertebrate cells.

15. Comparing Wnt and Fgf signalling gradient formation

15.1 Gradients through control and clearance

Wolpert's model of the French Flag suggests that the receiving cells are exposed to different concentrations of the morphogen to identify their position within a tissue. Therefore, the formation of concentration gradients is essential in tissue patterning. In addition to providing selected concentrations to the target cells via the mechanisms discussed above, the extracellular matrix or target cells can shape the gradient.

Extracellularly, enzymatic cleavage of the morphogens can lead to the inactivation of the signal, hence a lower signal concentration in the ECM at a distance to the source. Secreted inhibitors of Wnt proteins are also expressed by Wnt signalling; for example, a highly conserved extracellular protein, Notum, binds and deacetylates Wnts, which reduces their signalling activity (Kakugawa et al., 2015).

In Fgf signalling, the matrix metalloprotease Mmp2 in Drosophila can restrict Fgf signalling in the extracellular region (Wang et al., 2010). Mmp2 could control extracellular Fgf availability through the proteolytic cleavage of ECM or the FgfR. Similarly, the protease known as high-temperature requirement A1 (HtrA1) is essential for dorsoventral patterning during early zebrafish embryogenesis by functioning as an extracellular protease of Fgf signalling. Hence, HtrA1 controls the extracellular levels of available Fgf8, thus shaping the Fgf gradient (Kim et al., 2012). In addition, HtrA1 can modulate Fgf signalling by cleaving HSPGs in Xenopus (Hou et al., 2007).

16. Controlled morphogen transport shapes the gradient

In addition to extracellular morphogen degradation, a further intriguing possibility of generating a gradient is the modulation of the transport route. There is evidence in zebrafish that cells closer to the Wnt source are targeted more often by Wnt by cytonemes than cells further away (Stanganello et al., 2015). Spatially controlled contacts can lead to establishing the Wnt signalling gradient (Rosenbauer et al., 2020). Complementary, there is evidence in Drosophila that cells closer to the Fgf source receive higher levels of Fgf by cytonemes, and the higher Fgf signalling levels are required to induce more Fgf-receiving cytonemes to maintain the signalling levels (Du et al., 2018). Similarly, cells that are further away from the source lack long cytonemes to receive Fgf leading to suppression of cytoneme formation. The interplay of the positive and negative feedbacks in recipient tissue can shape the Fgf morphogen gradient in a target-specific manner (Fig. 3B). Further studies are required to explore in which contexts a morphogen degradation mechanism is favoured over other gradient-forming mechanisms.

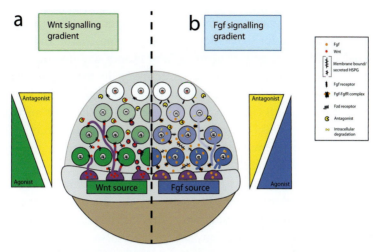

Fig. 3 Hypothetical model of Wnt and Fgf gradient formation in the zebrafish gastrula. (A) Wnt signalling utilises transport mechanisms such as exosomes and cytonemes, where the Wnt ligands are 'carried' to target cells in a concentration-dependent manner. In addition, antagonistic systems (i.e., Notum, sFRP) reduce extracellular ligand availability and intracellular antagonists (i.e., Axin2) inhibit signal activation. (B) Fgf signals utilise HSPGs and cytonemes to form a gradient. Furthermore, antagonistic systems (i.e., HtrA1) reduce extracellular ligand availability, and intracellular antagonists (i.e., Sprouty4) reduce signal activation.

17. Signalling modulators in the target cells

Finally, the target cell itself plays an essential role in gradient formation. The uptake and degradation of morphogens into the target cells, also known as the restrictive clearance model (RCM), provides a way to generate a gradient (Scholpp & Brand, 2004). This process ensures clearances of the Fgf-FgfR complex and can control the formation of the morphogen gradient. One proposed mechanism for this is Fgf-FgfR internalisation through endocytosis. Studies within zebrafish have demonstrated that Fgf8 is removed from the ECM by cells and quickly degraded via lysosomes. If Rab5-mediated endocytosis is inhibited, there is an increase in Fgf8 signalling intensity and range (Scholpp & Brand, 2004; Nowak et al., 2011). A popular receptor tyrosine kinase (RTK) endocytosis mechanism is through the budding of RTK-containing clathrin-coated pits to form intracellular vesicles. These vesicles fuse with early endosomes, which undergo further transformation to reach the lysosome for degradation (Rengarajan et al., 2014; Yuan & Song, 2020). It is unclear how RCM impacts on the Wnt gradient, as endocytosis can remove the ligand, however, it is also needed for signalling (Brunt & Scholpp, 2018). In contrast to Fgf signalling, the β-catenin-dependent Wnt ligands require endocytosis to activate signalling and regulate gene transcription in the responding cells. However, Wnts can also be removed from the Lrp6 signalosome by uptake and degradation. The RCM can also be expanded to the Wnt antagonists, and the removal of Dkk1 by receptor-mediated endocytosis further complicates the analysis (Kawamura et al., 2020). In conclusion, the role of endocytosis in Wnt signalling needs to be clarified.

In addition, intracellular signalling antagonists influence the gradient. For example, the control of the Fgf gradient is also thought to be regulated by protein modulators that antagonise the Fgf ligands rendering them inactive and unable to bind to any additional FgfRs. An example is the protein Sprouty (Spry). It has been shown that Spry inhibits tracheal branching during embryogenesis within Drosophila. Within Spry mutants, prestalk cells located further from Fgf/Bnl signalling were induced through Fgf to bud-like normal branch cells. Spry has been found to interrupt signalling conduction between the FgfR and the intracellular Ras protein, which is required for MAPK signalling (Casci et al., 1999). There has been the identification of another Fgf modulator within Xenopus and zebrafish. Overexpression of the protein encoded by *sef* causes a decrease in Fgf signalling while impairing *sef* leads to overstimulation of the pathway. Sef was shown to be a transmembrane protein that binds to FgfRs via a Tyr39 intracellular domain on Sef (Tsang et al., 2002).

The most famous example of a negative-feedback loop regulator in the Wnt signalling pathway is Axin2. Axin2 restrains Wnt signalling activity by limiting the amounts of β-catenin in the receiving cell (Lustig et al., 2002). The upregulation of Axin2 appears to be an event in β-catenin/Wnt-induced signalling and is initiated within 30 min (Lustig et al., 2002). Axin2 is one of the most potent antagonists and can specifically block Wnt/β-catenin signalling when required (Nusse & Clevers, 2017).

18. Final remarks on the importance of transport modes in morphogen gradient formation

The evidence detailed above indicates that there are multiple methods for establishing a morphogen gradient, and the mechanism used for transporting morphogens strongly influences the gradient. The post-translational palmitoleation of Wnt suggests that the transfer of Wnts requires interaction with glypicans or other binding proteins, which may facilitate transport between directly adjacent cells. Further, the discovery of membrane-bound Wnt upon cytonemes or EVs allows for their long-range transport without diffusion. The combination of this evidence supports the theory that Wnt is a short-range signalling molecule that can act in a long-range fashion via its presentation through membrane-bound carriers. It will be fascinating to study if the cytoneme plasma membranes need to fulfil certain characteristics in their composition of transmembrane proteins, lipids and glycocalyx to allow Wnt transport. Furthermore, by comparing a diffusion-based transport to a cytoneme-mediated transport mechanism and the temporal emergence of tissue pattern, mathematical modelling suggests that cytoneme-based transport of morphogens leads to an early establishment of sufficiently different concentrations to assign different cellular fates within a tissue. Diffusion-based transport also allows for the establishment of a pattern, but this takes significantly more time to establish the correct cell fate areas in a tissue (Rosenbauer et al., 2020). It will be essential to correlate the timing of the establishment of a pattern to the dissemination mode of the morphogen.

Vertebrate Fgf ligands, on the other hand, are considered to have the ability to diffuse passively, limiting it to producing a short-range gradient, as long-range distances will impede the speed and efficiency of the signalling. Transport and signalling require strong interactions with HSPGs. The fast and short-range nature of Fgf diffusion requires tight regulation to ensure the precise formation of a morphogen gradient. This explains the requirement of mechanisms to mediate controlled clearance of Fgf from the ECM, such as restricted clearance via endocytosis. In Drosophila, there is evidence that the FgfR can be loaded on cytonemes to pick up the ligand at a distance. Recently, it has been shown that also the Fgf ligands like Bnl and Pyr are membrane-tethered and directed, long-range transport from Fgfs via cytonemes. The idea of short-range signalling via diffusible signals and long-range signalling via a cytoneme/EV-based transport is intriguing. In support of this idea, experiments in Drosophila suggest that removing the membrane anchor of Fgf/Bnl reduces the long-range dispersion and signalling (Du et al., 2022). Consistently, GPI-tethered Fgf/Bnl increases the range of signalling and promotes tracheation as it can be loaded on cytonemes.

It is obvious that both Wnt and Fgf signalling play pivotal roles within embryogenesis and are essential for correct development. This review highlights the importance of each of their modes of transportation to create the desired morphogen gradient – a prerequisite of proper body patterning. Furthermore, combining these transport mechanisms allows them to reach and interact with multiple cell types, directing targeted cell differentiation and controlling gastrulation. Although it is clear that the accuracy and precision of ligand presentation to its receptor primarily rely upon these transportation mechanisms, the context behind which mechanism is used is currently unknown. Further research must be completed to reveal the spatiotemporal processes underlying Wnt and Fgf distribution to understand their role in embryonic development.

Acknowledgement

E.J.C. is supported by a studentship from the BBSRC Southwest Biosciences Doctoral Training Partnership (SWBio DTP3; BB/T008741/1). Research in the Scholpp lab is supported by the BBSRC (Research Grants, BB/S016295/1, BB/X008401/1) and by the Living Systems Institute, University of Exeter. In addition, we want to thank Drs Lucy Brunt (University of Exeter, UK) and Sougata Roy (University of Maryland, US) for the images on cytonemes in Fig. 2. Finally, we thank Sougata Roy, David Virshup (DUKE-NUS, Singapore), and Benjamin Housden (LSI Exeter) for their critical comments on the manuscript.

References

Alvarez, I. S., Araujo, M. A., & Nieto, M. A. (1998). Neural induction in whole chick embryo cultures by FGF. *Developmental Biology, 199*, 42–54.

Alvarez-Rodrigo, I., Willnow, D., & Vincent, J.-P. (2023). The logistics of Wnt production and delivery. *Current Topics in Developmental Biology.* Elsevier, 1–60.

Andre, P., Song, H., Kim, W., Kispert, A., & Yang, Y. (2015). Wnt5a and Wnt11 regulate mammalian anterior-posterior axis elongation. *Development (Cambridge, England), 142*, 1516–1527.

Baker, J. C., Beddington, R. S., & Harland, R. M. (1999). Wnt signaling in Xenopus embryos inhibits bmp4 expression and activates neural. *Genes & Development, 13*, 3149–3159.

Bouwmeester, T., Kim, S.-H., Sasai, Y., Lu, B., & Robertis, E. M. D. (1996). Cerberus is a head-inducing secreted factor expressed in the anterior endoderm of Spemann's organizer. *Nature, 382*, 595–601.

Brafman, D., & Willert, K. (2017). Wnt/β-catenin signaling during early vertebrate neural development. *Developmental Neurobiology, 77*, 1239–1259.

Briscoe, J., & Small, S. (2015). Morphogen rules: Design principles of gradient-mediated embryo patterning. *Development (Cambridge, England), 142*, 3996–4009.

Brunt, L., & Scholpp, S. (2018). The function of endocytosis in Wnt signaling. *Cellular and Molecular Life Sciences, 75*, 785–795.

Brunt, L., Greicius, G., Rogers, S., Evans, B. D., Virshup, D. M., Wedgwood, K. C., & Scholpp, S. (2021). Vangl2 promotes the formation of long cytonemes to enable distant Wnt/β-catenin signaling. *Nature Communications, 12*, 2058.

Casci, T., Vinós, J., & Freeman, M. (1999). Sprouty, an intracellular inhibitor of Ras signaling. *Cell, 96*, 655–665.

Christian, J. L., & Moon, R. T. (1993). Interactions between Xwnt-8 and Spemann organizer signaling pathways generate dorsoventral pattern in the embryonic mesoderm of Xenopus. *Genes & Development, 7*, 13–28.

Ciruna, B., & Rossant, J. (2001). FGF signaling regulates mesoderm cell fate specification and morphogenetic movement at the primitive streak. *Developmental Cell, 1*, 37–49.

Crossley, P. H., Martinez, S., & Martin, G. R. (1996). Midbrain development induced by FGF8 in the chick embryo. *Nature, 380*, 66–68.

De Robertis, E. M., & Kuroda, H. (2004). Dorsal-ventral patterning and neural induction in Xenopus embryos. *Annual Review of Cell and Developmental Biology, 20*, 285–308.

Delaune, E., Lemaire, P., & Kodjabachian, L. (2005). Neural induction in Xenopus requires early FGF signalling in addition to BMP inhibition. *Development, 132*(2), 299–310.

Ding, Y., Ploper, D., Sosa, E. A., Colozza, G., Moriyama, Y., Benitez, M. D., ... De Robertis, E. M. (2017). Spemann organizer transcriptome induction by early beta-catenin, Wnt, Nodal, and Siamois signals in Xenopus laevis. *Proceedings of the National Academy of Sciences, 114*, E3081–E3090.

Du, L., Sohr, A., Yan, G., & Roy, S. (2018). Feedback regulation of cytoneme-mediated transport shapes a tissue-specific FGF morphogen gradient. *Elife, 7*, e38137.

Du, L., Sohr, A., Li, Y., & Roy, S. (2022). GPI-anchored FGF directs cytoneme-mediated bidirectional contacts to regulate its tissue-specific dispersion. *Nature Communications, 13*, 1–19.

Działo, E., Rudnik, M., Koning, R. I., Czepiel, M., Tkacz, K., Baj-Krzyworzeka, M., ... Błyszczuk, P. (2019). WNT3a and WNT5a transported by exosomes activate WNT signaling pathways in human cardiac fibroblasts. *International Journal of Molecular Sciences, 20*, 1436.

Flaumenhaft, R., Moscatelli, D., & Rifkin, D. B. (1990). Heparin and heparan sulfate increase the radius of diffusion and action of basic fibroblast growth factor. *The Journal of Cell Biology, 111*, 1651–1659.

Franch-Marro, X., Marchand, O., Piddini, E., Ricardo, S., Alexandre, C., & Vincent, J.-P. (2005). Glypicans shunt the Wingless signal between local signalling and further transport. *Development, 132*(4), 659–666.

Glinka, A., Wu, W., Delius, H., Monaghan, A. P., Blumenstock, C., & Niehrs, C. (1998). Dickkopf-1 is a member of a new family of secreted proteins and functions in head induction. *Nature, 391*, 357–362.

Gospodarowicz, D., & Cheng, J. (1986). Heparin protects basic and acidic FGF from inactivation. *Journal of Cellular Physiology, 128*, 475–484.

Greco, V., Hannus, M., & Eaton, S. (2001). Argosomes: A potential vehicle for the spread of morphogens through epithelia. *Cell, 106*, 633–645.

Gross, J. C., Chaudhary, V., Bartscherer, K., & Boutros, M. (2012). Active Wnt proteins are secreted on exosomes. *Nature Cell Biology, 14*, 1036–1045.

Guan, J.-L., Machamer, C. E., & Rose, J. K. (1985). Glycosylation allows cell-surface transport of an anchored secretory protein. *Cell, 42*, 489–496.

Han, C., Yan, D., Belenkaya, T. Y., & Lin, X. (2005). Drosophila glypicans Dally and Dally-like shape the extracellular Wingless morphogen gradient in the wing disc. *Development, 132*(4), 667–679.

Harada, M., Murakami, H., Okawa, A., Okimoto, N., Hiraoka, S., Nakahara, T., ... Mizutani-Koseki, Y. (2009). FGF9 monomer–dimer equilibrium regulates extracellular matrix affinity and tissue diffusion. *Nature Genetics, 41*, 289–298.

Hébert, J. M., & Fishell, G. (2008). The genetics of early telencephalon patterning: Some assembly required. *Nature Reviews. Neuroscience, 9*, 678–685.

Heikinheimo, M., Lawshé, A., Shackleford, G. M., Wilson, D. B., & MacArthur, C. A. (1994). Fgf-8 expression in the post-gastrulation mouse suggests roles in the development of the face, limbs and central nervous system. *Mechanisms of Development, 48*, 129–138.

Heisenberg, C.-P., Houart, C., Take-uchi, M., Rauch, G.-J., Young, N., Coutinho, P., ... Geisler, R. (2001). A mutation in the Gsk3-binding domain of zebrafish Masterblind/Axin1 leads to a fate transformation of telencephalon and eyes to diencephalon. *Genes & Development, 15*, 1427–1434.

Hongo, I., & Okamoto, H. (2020). Fgf/Ets signalling in Xenopus ectoderm initiates neural induction and patterning in an autonomous and paracrine manners. bioRxiv.

Hongo, I., & Okamoto, H. (2022). FGF/MAPK/Ets signaling in Xenopus ectoderm contributes to neural induction and patterning in an autonomous and paracrine manner, respectively. *Cells & Development, 170*, 203769.

Hou, S., Maccarana, M., Min, T. H., Strate, I., & Pera, E. M. (2007). The secreted serine protease xHtrA1 stimulates long-range FGF signaling in the early Xenopus embryo. *Developmental Cell, 13*, 226–241.

Houart, C., Westerfield, M., & Wilson, S. W. (2001). A small population of anterior cells patterns the forebrain during zebrafish gastrulation. *Nature, 391*(6669), 788–792.

Houart, C., Caneparo, L., Heisenberg, C.-P., Barth, K. A., Take-Uchi, M., & Wilson, S. W. (2002). Establishment of the telencephalon during gastrulation by local antagonism of Wnt signaling. *Neuron, 35*, 255–265.

Huang, H., & Kornberg, T. B. (2015). Myoblast cytonemes mediate Wg signaling from the wing imaginal disc and Delta-Notch signaling to the air sac primordium. *eLife, 4*, e06114.

Huang, H., Liu, S., & Kornberg, T. B. (2019). Glutamate signaling at cytoneme synapses. *Science (New York, N. Y.), 363*, 948–955.

Janda, C. Y., Waghray, D., Levin, A. M., Thomas, C., & Garcia, K. C. (2012). Structural basis of Wnt recognition by Frizzled. *Science (New York, N. Y.), 337*, 59–64.

Kaiser, K., Gyllborg, D., Procházka, J., Salašová, A., Kompaníková, P., Molina, F. L., ... Procházková, M. (2019). WNT5A is transported via lipoprotein particles in the cerebrospinal fluid to regulate hindbrain morphogenesis. *Nature Communications, 10*, 1–15.

Kakugawa, S., Langton, P. F., Zebisch, M., Howell, S. A., Chang, T.-H., Liu, Y., ... Snijders, A. P. (2015). Notum deacylates Wnt proteins to suppress signalling activity. *Nature, 519*, 187–192.

Kalluri, R., & Weinberg, R. A. (2009). The basics of epithelial-mesenchymal transition. *The Journal of Clinical Investigation, 119*, 1420–1428.

Kawamura, N., Takaoka, K., Hamada, H., Hadjantonakis, A.-K., Sun-Wada, G.-H., & Wada, Y. (2020). Rab7-mediated endocytosis establishes patterning of Wnt activity through inactivation of Dkk antagonism. *Cell Reports, 31*, 107733.

Kengaku, M., & Okamoto, H. (1993). Basic fibroblast growth factor induces differentiation of neural tube and neural crest lineages of cultured ectoderm cells from Xenopus gastrula. *Development (Cambridge, England), 119*, 1067–1078.

Khokha, M. K., Yeh, J., Grammer, T. C., & Harland, R. M. (2005). Depletion of three BMP antagonists from Spemann's organizer leads to a catastrophic loss of dorsal structures. *Developmental Cell, 8*, 401–411.

Kiecker, C., & Niehrs, C. (2001). A morphogen gradient of Wnt/β-catenin signalling regulates anteroposterior neural patterning in Xenopus.

Kim, G.-Y., Kim, H.-Y., Kim, H.-T., Moon, J.-M., Kim, C.-H., Kang, S., & Rhim, H. (2012). HtrA1 is a novel antagonist controlling fibroblast growth factor (FGF) signaling via cleavage of FGF8. *Molecular and Cellular Biology, 32*, 4482–4492.

Korkut, C., Ataman, B., Ramachandran, P., Ashley, J., Barria, R., Gherbesi, N., & Budnik, V. (2009). Trans-synaptic transmission of vesicular Wnt signals through Evi/Wntless. *Cell, 139*, 393–404.

Kornberg, T. B., & Roy, S. (2014). Cytonemes as specialized signaling filopodia. *Development (Cambridge, England), 141*, 729–736.

Kudoh, T., Concha, M. L., Houart, C., Dawid, I. B., & Wilson, S. W. (2004). Combinatorial Fgf and Bmp signalling patterns the gastrula ectoderm into prospective neural and epidermal domains. *Development, 131*(15), 3581–3592. https://doi.org/10.1242/dev.01227.

LaBonne, C., Burke, B., & Whitman, M. (1995). Role of MAP kinase in mesoderm induction and axial patterning during Xenopus development. *Development (Cambridge, England), 121*, 1475–1486.

Lamb, T. M., & Harland, R. M. (1995). Fibroblast growth factor is a direct neural inducer, which combined with noggin generates anterior-posterior neural pattern. *Development (Cambridge, England), 121*, 3627–3636.

Lamb, T. M., Knecht, A. K., Smith, W. C., Stachel, S. E., Economides, A. N., Stahl, N., ... Harland, R. M. (1993). Neural induction by the secreted polypeptide noggin. *Science (New York, N. Y.), 262*, 713–718.

Lekven, A. C., Thorpe, C. J., Waxman, J. S., & Moon, R. T. (2001). Zebrafish wnt8 encodes two wnt8 proteins on a bicistronic transcript and is required for mesoderm and neurectoderm patterning. *Developmental Sell, 1*, 103–114.

Leyns, L., Bouwmeester, T., Kim, S.-H., Piccolo, S., & De Robertis, E. M. (1997). Frzb-1 is a secreted antagonist of Wnt signaling expressed in the Spemann organizer. *Cell, 88*, 747–756.

Lin, X. (2004). Functions of heparan sulfate proteoglycans in cell signaling during development, *Development, 131*(24), 6009–6021.

Luga, V., Zhang, L., Viloria-Petit, A. M., Ogunjimi, A. A., Inanlou, M. R., Chiu, E., ... Wrana, J. L. (2012). Exosomes mediate stromal mobilization of autocrine Wnt-PCP signaling in breast cancer cell migration. *Cell, 151*, 1542–1556.

Lustig, B., Jerchow, B., Sachs, M., Weiler, S., Pietsch, T., Karsten, U., ... Birchmeier, W. (2002). Negative feedback loop of Wnt signaling through upregulation of conductin/axin2 in colorectal and liver tumors. *Molecular and Cellular Biology, 22*, 1184–1193.

MacNicol, A. M., Muslin, A. J., & Williams, L. T. (1993). Raf-1 kinase is essential for early Xenopus development and mediates the induction of mesoderm by FGF. *Cell, 73*, 571–583.

Mattes, B., Dang, Y., Greicius, G., Kaufmann, L. T., Prunsche, B., Rosenbauer, J., ... Nienhaus, G. U. (2018). Wnt/PCP controls spreading of Wnt/β-catenin signals by cytonemes in vertebrates. *Elife, 7*, e36953.

McGough, I. J., Vecchia, L., Bishop, B., Malinauskas, T., Beckett, K., Joshi, D., ... Vincent, J.-P. (2020). Glypicans shield the Wnt lipid moiety to enable signalling at a distance. *Nature, 585*, 85–90.

Mehta, S., Hingole, S., & Chaudhary, V. (2021). The emerging mechanisms of Wnt secretion and signaling in development. *Frontiers in Cell and Developmental Biology, 2191*.

Mihara, E., Hirai, H., Yamamoto, H., Tamura-Kawakami, K., Matano, M., Kikuchi, A., ... Takagi, J. (2016). Active and water-soluble form of lipidated Wnt protein is maintained by a serum glycoprotein afamin/α-albumin. *Elife, 5*, e11621.

Mii, Y., & Taira, M. (2009). Secreted Frizzled-related proteins enhance the diffusion of Wnt ligands and expand their signalling range. *Development (Cambridge, England), 136*, 4083–4088.

Mii, Y., Yamamoto, T., Takada, R., Mizumoto, S., Matsuyama, M., Yamada, S., ... Taira, M. (2017). Roles of two types of heparan sulfate clusters in Wnt distribution and signaling in Xenopus. *Nature Communications, 8*, 1973.

Mohammadi, M., Olsen, S. K., & Ibrahimi, O. A. (2005). Structural basis for fibroblast growth factor receptor activation. *Cytokine & Growth Factor Reviews, 16*, 107–137.

Mulligan, K. A., Fuerer, C., Ching, W., Fish, M., Willert, K., & Nusse, R. (2012). Secreted Wingless-interacting molecule (Swim) promotes long-range signaling by maintaining Wingless solubility. *Proceedings of the National Academy of Sciences, 109*, 370–377.

Niehrs, C. (2012). The complex world of WNT receptor signalling. *Nature Reviews. Molecular Cell Biology, 13*, 767–779.

Nordström, U., Jessell, T. M., & Edlund, T. (2002). Progressive induction of caudal neural character by graded Wnt signaling. *Nature Neuroscience, 5*, 525–532.

Nowak, M., Machate, A., Yu, S. R., Gupta, M., & Brand, M. (2011). Interpretation of the FGF8 morphogen gradient is regulated by endocytic trafficking. *Nature Cell Biology, 13*, 153–158.

Nusse, R., & Varmus, H. E. (1982). Many tumors induced by the mouse mammary tumor virus contain a provirus integrated in the same region of the host genome. *Cell, 31*, 99–109.

Nusse, R., & Clevers, H. (2017). Wnt/β-catenin signaling, disease, and emerging therapeutic modalities. *Cell, 169*, 985–999.

Nüsslein-Volhard, C., & Wieschaus, E. (1980). Mutations affecting segment number and polarity in Drosophila. *Nature, 287*, 795–801.

Nygaard, R., Yu, J., Kim, J., Ross, D. R., Parisi, G., Clarke, O. B., ... Mancia, F. (2021). Structural basis of WLS/Evi-mediated Wnt transport and secretion. *Cell, 184*, 194–206.e114.

Ornitz, D. M. (2000). FGFs, heparan sulfate and FGFRs: Complex interactions essential for development. *Bioessays: News and Reviews in Molecular, Cellular and Developmental Biology, 22*, 108–112.

Ornitz, D. M., & Itoh, N. (2015). The fibroblast growth factor signaling pathway. *Wiley Interdisciplinary Reviews: Developmental Biology, 4*, 215–266.

Ota, S., Tonou-Fujimori, N., & Yamasu, K. (2009). The roles of the FGF signal in zebrafish embryos analyzed using constitutive activation and dominant-negative suppression of different FGF receptors. *Mechanisms of Development, 126*, 1–17.

Panáková, D., Sprong, H., Marois, E., Thiele, C., & Eaton, S. (2005). Lipoprotein particles are required for Hedgehog and Wingless signalling. *Nature, 435*, 58–65.

Parr, B. A., Shea, M. J., Vassileva, G., & McMahon, A. P. (1993). Mouse Wnt genes exhibit discrete domains of expression in the early embryonic CNS and limb buds. *Development (Cambridge, England), 119*, 247–261.

Patel, A., Wu, Y., Han, X., Su, Y., Maugel, T., Shroff, H., & Roy, S. (2022). Cytonemes coordinate asymmetric signaling and organization in the Drosophila muscle progenitor niche. *Nature Communications, 13*, 1185.

Pera, E. M., Ikeda, A., Eivers, E., & De Robertis, E. M. (2003). Integration of IGF, FGF, and anti-BMP signals via Smad1 phosphorylation in neural induction. *Genes & Development, 17*, 3023–3028.

Petit, I., Levy, A., Estrach, S., Féral, C. C., Trentin, A. G., Dingli, F., ... Théry, C. (2022). Fibroblast growth factor-2 bound to specific dermal fibroblast-derived extracellular vesicles is protected from degradation. *Scientific Reports, 12*, 22131.

Piccolo, S., Agius, E., Leyns, L., Bhattacharyya, S., Grunz, H., Bouwmeester, T., & Robertis, E. D. (1999). The head inducer Cerberus is a multifunctional antagonist of Nodal, BMP and Wnt signals. *Nature, 397*, 707–710.

Pownall, M. E., Tucker, A. S., Slack, J. M., & Isaacs, H. V. (1996). eFGF, Xcad3 and Hox genes form a molecular pathway that establishes the anteroposterior axis in Xenopus. *Development (Cambridge, England), 122*, 3881–3892.

Proffitt, K. D., & Virshup, D. M. (2012). Precise regulation of porcupine activity is required for physiological Wnt signaling. *Journal of Biological Chemistry, 287*, 34167–34178.

Ramírez-Weber, F.-A., & Kornberg, T. B. (1999). Cytonemes: Cellular processes that project to the principal signaling center in Drosophila imaginal discs. *Cell, 97*, 599–607.

Rapraeger, A. C., Krufka, A., & Olwin, B. B. (1991). Requirement of heparan sulfate for bFGF-mediated fibroblast growth and myoblast differentiation. *Science (New York, N. Y.), 252*, 1705–1708.

Reifers, F., Böhli, H., Walsh, E. C., Crossley, P. H., Stainier, D. Y., & Brand, M. (1998). Fgf8 is mutated in zebrafish acerebellar (ace) mutants and is required for maintenance of midbrain-hindbrain boundary development and somitogenesis. *Development, 125*, 2381–2395.

Rengarajan, C., Matzke, A., Reiner, L., Orian-Rousseau, V., & Scholpp, S. (2014). Endocytosis of Fgf8 is a double-stage process and regulates spreading and signaling. *PLoS One, 9*, e86373.

Ribisi, S. Jr, Mariani, F. V., Aamar, E., Lamb, T. M., Frank, D., & Harland, R. M. (2000). Ras-mediated FGF signaling is required for the formation of posterior but not anterior neural tissue in Xenopus laevis. *Developmental Biology, 227*, 183–196.

Rogers, S., Zhang, C., Anagnostidis, V., Liddle, C., Fishel, M. I., Gielen, F., & Scholpp, S. (2023). Cancer-associated fibroblasts influence Wnt/PCP signaling in gastric cancer cells by cytoneme-based dissemination of ROR2. *Proceedings National Academy Sciences, 120*(39), e2217612120.

Rosenbauer, J., Zhang, C., Mattes, B., Reinartz, I., Wedgwood, K., Schindler, S., ... Schug, A. (2020). Modeling of Wnt-mediated tissue patterning in vertebrate embryogenesis. *PLoS Computational Biology, 16*, e1007417.

Routledge, D., Rogers, S., Ono, Y., Brunt, L., Meniel, V., Tornillo, G., ... Scholpp, S. (2022). The scaffolding protein flot2 promotes cytoneme-based transport of wnt3 in gastric cancer. *Elife, 11*, e77376.

Roy, S., Hsiung, F., & Kornberg, T. B. (2011). Specificity of Drosophila cytonemes for distinct signaling pathways. *Science (New York, N. Y.), 332*, 354–358.

Roy, S., Huang, H., Liu, S., & Kornberg, T. B. (2014). Cytoneme-mediated contact-dependent transport of the Drosophila decapentaplegic signaling protein. *Science (New York, N. Y.), 343*, 1244624.

Sasai, Y., Lu, B., Steinbeisser, H., & De Robertis, E. M. (1995). Regulation of neural induction by the Chd and Bmp-4 antagonistic patterning signals in Xenopus. *Nature, 376*, 333–336.

Sato, T., Araki, I., & Nakamura, H. (2001). Inductive signal and tissue responsiveness defining the tectum and the cerebellum. *Development (Cambridge, England), 128*(13), 2461–2469.

Scholpp, S., & Brand, M. (2004). Endocytosis controls spreading and effective signaling range of Fgf8 protein. *Current Biology, 14*(20), 1834–1841.

Shimizu, T., Bae, Y.-K., Muraoka, O., & Hibi, M. (2005). Interaction of Wnt and caudal-related genes in zebrafish posterior body formation. *Developmental Biology, 279*, 125–141.

Slack, J., Isaacs, H., Song, J., Durbin, L., & Pownall, M. (1996). The role of fibroblast growth factors in early Xenopus development. *Biochemical Society Symposium*, 1–12.

Spemann, H., & Mangold, H. (1924). über Induktion von Embryonalanlagen durch Implantation artfremder Organisatoren. *Archiv für mikroskopische Anatomie und Entwicklungsmechanik, 100*, 599–638.

Stanganello, E., & Scholpp, S. (2016). Role of cytonemes in Wnt transport. *Journal of Cell Science, 129*, 665–672.

Stanganello, E., Hagemann, A. I., Mattes, B., Sinner, C., Meyen, D., Weber, S., ... Scholpp, S. (2015). Filopodia-based Wnt transport during vertebrate tissue patterning. *Nature Communications, 6*(1), 14.

Streit, A., Berliner, A. J., Papanayotou, C., Sirulnik, A., & Stern, C. D. (2000). Initiation of neural induction by FGF signalling before gastrulation. *Nature, 406*, 74–78.

Streit, A., Lee, K. J., Woo, I., Roberts, C., Jessell, T. M., & Stern, C. D. (1998). Chordin regulates primitive streak development and the stability of induced neural cells, but is not sufficient for neural induction in the chick embryo. *Development (Cambridge, England), 125*, 507–519.

Sun, X., Meyers, E. N., Lewandoski, M., & Martin, G. R. (1999). Targeted disruption of Fgf8 causes failure of cell migration in the gastrulating mouse embryo. *Genes & Development, 13*, 1834–1846.

Takada, S., Stark, K. L., Shea, M. J., Vassileva, G., McMahon, J. A., & McMahon, A. P. (1994). Wnt-3a regulates somite and tailbud formation in the mouse embryo. *Genes & Development, 8*, 174–189.

Teven, C. M., Farina, E. M., Rivas, J., & Reid, R. R. (2014). Fibroblast growth factor (FGF) signaling in development and skeletal. *Genes & Diseases, 1*, 199–213.

Thiery, J. P., Acloque, H., Huang, R. Y., & Nieto, M. A. (2009). Epithelial-mesenchymal transitions in development and disease. *Cell, 139*, 871–890.

Tsang, M., Friesel, R., Kudoh, T., & Dawid, I. B. (2002). Identification of Sef, a novel modulator of FGF signalling. *Nature Cell Biology, 4*, 165–169.

Turing, A. M. (1952). The chemical basis of morphogenesis. *Bulletin of Mathematical Biology, 52*(1), 153–197.

Umbhauer, M., Marshall, C., Mason, C., Old, R., & Smith, J. (1995). Mesoderm induction in Xenopus caused by activation of MAP kinase. *Nature, 376*, 58–62.

Uren, A., Reichsman, F., Anest, V., Taylor, W. G., Muraiso, K., Bottaro, D. P., ... Rubin, J. S. (2000). Secreted frizzled-related protein-1 binds directly to Wingless and is a biphasic modulator of Wnt signaling. *Journal of Biological Chemistry, 275*, 4374–4382.

Waddington, C. H. (1940). *Organisers and genes*. Cambridge, England: The University Press, 160.

Wang, Q., Uhlirova, M., & Bohmann, D. (2010). Spatial restriction of FGF signaling by a matrix metalloprotease controls branching morphogenesis. *Developmental Cell, 18*, 157–164.

Wang, S., Krinks, M., Lin, K., Luyten, F. P., & Moos, M. Jr. (1997). Frzb, a secreted protein expressed in the Spemann organizer, binds and inhibits Wnt-8. *Cell, 88*, 757–766.

Wasylyk, B., Hagman, J., & Gutierrez-Hartmann, A. (1998). Ets transcription factors: Nuclear effectors of the Ras–MAP-kinase signaling pathway. *Trends in Biochemical Sciences, 23*, 213–216.

Wilson, S., Rydström, A., Trimborn, T., Willert, K., Nusse, R., Jessell, T. M., & Edlund, T. (2001). The status of Wnt signalling regulates neural and epidermal fates in the chick embryo. *Nature, 411*, 325–330.

Wolpert, L. (1969). Positional information and the spatial pattern of cellular differentiation. *Journal of Theoretical Biology, 25*, 1–47.

Xavier, C. P., Melikova, M., Chuman, Y., Üren, A., Baljinnyam, B., & Rubin, J. S. (2014). Secreted Frizzled-related protein potentiation versus inhibition of Wnt3a/β-catenin signaling. *Cellular Signalling, 26*, 94–101.

Yamaguchi, T. P. (2001). Heads or tails: Wnts and anterior–posterior patterning. *Current Biology, 11*, R713–R724.

Yamaguchi, T. P., Harpal, K., Henkemeyer, M., & Rossant, J. (1994). fgfr-1 is required for embryonic growth and mesodermal patterning during mouse gastrulation. *Genes & Development, 8*, 3032–3044.

Yan, D., & Lin, X. (2009). Shaping morphogen gradients by proteoglycans. *Cold Spring Harbor Perspectives in Biology, 1*, a002493.

Yanagi, T., Ito, K., Nishihara, A., Minamino, R., Mori, S., Sumida, M., & Hashimoto, C. (2015). The S pemann organizer meets the anterior-most neuroectoderm at the equator of early gastrulae in amphibian species. *Development, Growth & Differentiation, 57*, 218–231.

Yang, Y., & Mlodzik, M. (2015). Wnt-Frizzled/planar cell polarity signaling: Cellular orientation by facing the wind (Wnt). *Annual Review of Cell and Developmental Biology, 31*, 623–646.

Yayon, A., Klagsbrun, M., Esko, J. D., Leder, P., & Ornitz, D. M. (1991). Cell surface, heparin-like molecules are required for binding of basic fibroblast growth factor to its high affinity receptor. *Cell, 64*, 841–848.

Yu, S. R., Burkhardt, M., Nowak, M., Ries, J., Petrášek, Z., Scholpp, S., ... Brand, M. (2009). Fgf8 morphogen gradient forms by a source-sink mechanism with freely diffusing molecules. *Nature, 461*, 533–536.

Yuan, W., & Song, C. (2020). The emerging role of Rab5 in membrane receptor trafficking and signaling pathways. *Biochemistry Research International, 2020*.

Zhai, L., Chaturvedi, D., & Cumberledge, S. (2004). Drosophila wnt-1 undergoes a hydrophobic modification and is targeted to lipid rafts, a process that requires porcupine. *Journal of Biological Chemistry, 279*, 33220–33227.

Zhang, C., & Scholpp, S. (2019). Cytonemes in development. *Current Opinion in Genetics & Development, 57*, 25–30.

Zhang, C., Brunt, L., Ono, Y., Rogers, S., & Scholpp, S. (2023). Cytoneme-mediated transport of active Wnt5b–Ror2 complexes in zebrafish. *Nature*. https://doi.org/10.1038/s41586-023-06850-7.

Zhou, X., Sasaki, H., Lowe, L., Hogan, B. L., & Kuehn, M. R. (1993). Nodal is a novel TGF-β-like gene expressed in the mouse node during gastrulation. *Nature, 361*, 543–547.

Printed in the United States
by Baker & Taylor Publisher Services